Principles of cosmology and gravitation

The frontispiece (overleaf) shows types of galaxy
(*reading from top to bottom*)

Elliptical: left: NGC4406; right: NGC3115
Normal spiral: left: NGC3031; right: NGC5457
Barred spiral: left: NGC2217; right: NGC1300
Irregular: left: NGC3034; right: NGC3109

(*Photographs from the Hale Observatories*)

Principles of cosmology and gravitation

MICHAEL BERRY

H H Wills Physics Laboratory,
University of Bristol

Institute of Physics Publishing,
Bristol and Philadelphia

First published 1976 by Cambridge University Press
Reprinted 1978
First published 1989 by IOP Publishing Ltd
Reprinted 1991, 1993
British Library Cataloguing in Publication Data

Berry, Michael, *1941–*
  Principles of cosmology and gravitation.
  1. Astronomy. Cosmology
  I. Title
  523.1

ISBN 0-85274-037-9

Library of Congress Cataloging-in-Publication Data are available

This impression reproduced from the first edition type by kind permission of Cambridge University Press

Cover photograph of NGC 5457 reproduced courtesy of the Royal Astronomical Society

First published under the Adam Hilger imprint by IOP Publishing Ltd
Reprinted under the Institute of Physics Publishing imprint, the book imprint of the Institute of Physics, London, by IOP Publishing Ltd, Techno House, Redcliffe Way, Bristol BS1 6NX, UK

US Editorial Office: IOP Publishing Inc., The Public Ledger Building, Suite 1035, Independence Square, Philadelphia, PA19106

Printed in Great Britain at the University Press, Cambridge

Contents

Preface to the 1989 reprint

The decision to make this book available again, after it has been out of print for several years, was prompted by continuing requests from students and teachers. I have taken the opportunity to make many minor corrections, most of which were kindly supplied by readers.

Several new themes have entered the subject in the fifteen years since the original writing. The link between cosmology and particle physics has been strengthened by the development of inflationary theories, which provide an explanation for the high degree of uniformity of the early universe. Large-scale structure in the present universe has received extensive attention, partly because of the discovery of giant voids in which there are no galaxies and partly because of the application of fractal mathematics to describe hierarchical galaxy clustering. And the enthusiasm (in my view misplaced) for the anthropic principle reflects the revival of the idea that the evolution of the universe as a whole, and of us as a cosmologising species, are inevitably connected.

These are important developments, but I have not revised the book to incorporate them because I would not be able to do them justice. I hope that the original material will still serve as a useful introduction at the undergraduate level to gravitational relativity applied to the expanding universe – subjects that continue to form the backbone of cosmology.

Bristol 1988   *Michael Berry*

Preface to the first edition

Modern scientific cosmology is one of our grandest intellectual adventures. It is also physics, uninhibited, applied on the largest scale. Indeed, many people are first 'turned on' to physics by popular books or films about cosmology. What a pity, then, that the subject is rarely taught in universities. Perhaps this is because a suitable textbook is lacking. There are many advanced treatises for the specialist, and many elementary expositions for the lay reader, but at the undergraduate level there is a gap.

This book is designed to fill that gap, and so promote teaching of cosmology in universities. The aim is to describe the universe as revealed by observation, and to present a theoretical framework powerful enough to enable important cosmological formulae to be derived and numerical calculations performed.

Any serious treatment must grasp the nettle of Einstein's general theory of relativity, because this gives the best description of the behaviour of matter and light under the influence of gravity; it forms the basis of current 'standard cosmology', and is employed constantly in the interpretation of observations. Here we avoid an elaborate and formal discussion based on the tensor calculus. Of course it is necessary to introduce the general expression for the separation (or interval) between two events, and this involves the metric tensor of spacetime. However, it is possible in the case of the highly symmetrical spacetimes of elementary general relativity and cosmology to determine the metric tensor by employing Gauss's formula for the curvature of an ordinary two-dimensional surface instead of using the general Einstein field equations. The curvature of a surface is a concept that makes no demands on the credulity of a student, so that this approach is a convenient way to introduce the geometrical interpretation of gravity.

A previous exposure to the ideas of special relativity is assumed, as is a knowledge of calculus, including partial differentiation. This book is, therefore, a suitable text for the final year of an undergraduate physics course. Experience shows that the material can be covered comfortably in twenty-four lectures. Problems of varying difficulty are included, together with solutions.

In writing this book I have used a great variety of sources, and it is impossible to acknowledge them all. The works I found most helpful are included in the bibliography, as recommended additional reading. I am most grateful to Dr P. G. Drazin, Dr M. S. Longair and Professor J. F. Nye for critically reading the manuscript and correcting a number of errors (they are, of course, not responsible for any that remain).

Finally, I would like to thank my students for their gentle responsiveness to this introduction of cosmology into their curriculum. It will not help them get a job, nor will it help them serve the military–industrial complex or increase the gross national product. But it will, I hope, contribute to the revival of the old idea that physics should be, above all, 'natural philosophy'.

Bristol 1974 *Michael Berry*

1 *Introduction*

It is customary to start with definitions, which are often all too glib and rob a subject of its richness. Nevertheless it can help to fix our ideas if we do have a definition, providing we do not take it too seriously. According to *Chambers's Dictionary*, cosmology is 'the science of the universe as a whole'. Again, it is customary to divide a subject into neatly-separated sections, even though this can obscure the richness of their interconnections. Nevertheless, it does help to distinguish three main aspects of cosmology.

Firstly, we have *cosmography*: cataloguing the objects in the universe and charting their positions and motions. Unlike geographers, we are restricted to one vantage-point – the Earth – where we sit and receive electromagnetic radiation. All our information about the universe is contained in the directional distribution of this radiation (a star here, a galaxy there) and in its spectral composition (light, X-ray, radio, etc.). By comparison, we have learned very little from the analysis of cosmic rays and meteorites (objects falling in from space), or from our first toddling steps outside the Earth.

Secondly, there is *theoretical cosmology*, where we search for a framework within which to comprehend the information from cosmography. Even here the tidy scheme breaks down, because it is not possible to discuss even the simplest observations without a theoretical framework – for example, 'the distance of an object' can have at least five different meanings, depending on how it is measured. Theoretical cosmology employs the physical laws established on and near the Earth, and makes the outrageous extrapolation that they apply throughout the universe. But physics, even extrapolated, is not enough; to escape from the prison of our single vantage-point we need something more: a 'cosmological principle'. This is essentially philosophical in nature; it does not follow from the laws of physics. In simple terms, the cosmological principle says: 'There is nothing special, cosmologically, about the Earth; therefore our large-scale observations are the same as those which would be made by observers anywhere else in the universe'. How fickle our intellect is! To mediaeval man it was completely natural to consider the Earth as being at the centre of the universe, yet here we are, a mere few centuries later, elevating anti-anthropocentrism to the level of a basic principle. The cosmological principle is immensely powerful: it enables us to select from all the complicated solutions of the equations of physics those which have certain simple symmetries.

What kind of physics does cosmology require? I am afraid it is a pretty rich stew, whose basic ingredient is a theory of *gravitation*, since that is the dominant force on the cosmic scale. The best description of gravity that we have is Einstein's 'general theory of relativity', and this will form the core of our account of cosmology. To flavour the stew there will be a bit of electromagnetism, some thermodynamics, and even a dash of particle physics. The proof of any stew lies in the eating, which in this case means comparison with observation. We shall see that theoretical cosmology based on general relativity is capable of explaining the observations. However, these are not yet precise enough or extensive enough to indicate which of a range of 'universe models' applies to our actual world.

The third aspect of cosmology is *cosmogony*; this is the study of the *origin* (or perhaps the infinitely distant past) of the universe. Here our arrogance will be extreme, for we shall extrapolate the laws of physics to the most distant times as well as places. We shall find ourselves interpreting the most modern radio-astronomical observations as giving detailed information about conditions in the chaos of a 'big bang' ten thousand million years ago. Distinguishing the distant past from the distant future involves the nature of time itself, especially its reversibility, and this leads to connections – still mysterious – between cosmology and laboratory physics.

2 Cosmography

2.1 What the universe contains

On the very largest scale, the universe consists of clusters of *galaxies*. About 10^{11} galaxies can be seen in the largest optical telescopes. From a cosmological point of view galaxies are the 'atoms' of the universe, and their distribution, motion and origin must be determined and explained. However, galaxies are of course complicated gravitationally-bound structures, each consisting of up to 10^{11} stars as well as gas clouds. These are often distributed throughout a disc-shaped region with a central core and spiral arms; however, many other forms are possible (see the frontispiece), and these are beautifully illustrated in the *Hubble atlas of galaxies* (see the bibliography). Each of the component stars is a nuclear powerhouse whose detailed behaviour is the concern of astrophysics. Our Sun is a typical star, situated about halfway out towards the edge of the disc of our galaxy. The projection of the plane of our galaxy can be seen in the sky as the faint white stripe called the 'Milky Way'. The Earth is one of a few planets gravitationally bound to the Sun. It is not known what fraction of stars have planetary systems, because no planets have been observed outside our solar system (although several have been inferred) and because the mechanism of planet formation is not understood. However, even if the mechanism relies on an improbable event like the near-collision of two stars (although this is now thought unlikely) the vast numbers of stars make it virtually certain that there are many planets in the universe not too dissimilar from our own. Therefore we can hardly claim our vantage-point to be exceptional; on the contrary, it is likely to be typical.

What about distances? It used to be common to illustrate the immensity and emptiness of space with imaginary scale models. For example, if the Sun were a watermelon in Piccadilly Circus, the Earth would be a grape pip a hundred metres away, and the nearest star would be another watermelon in Australia. Our imagination rapidly loses its grip on such models,

Table 1

Mean Earth–Moon distance $\equiv r_{\mathbb{C}} = 3.84 \times 10^8$ m $= 1.28$ light-seconds

Mean Earth–Sun distance $\quad \equiv r_{\odot} = 1.496 \times 10^{11}$ m $= 8.3$ light-minutes

$\qquad\qquad\qquad\qquad\quad 9.46 \times 10^{15}$ m $= 1$ light-year $= 0.307$ pc

$\qquad\qquad\qquad\qquad\quad 3.26$ light years $= 1$ pc

Distance to nearest star $\qquad \approx 4$ light years ≈ 1.2 pc

Diameter of our galaxy $\qquad \approx 10^5$ light-years $\approx 3 \times 10^4$ pc

Distance of nearest large galaxy $\approx 2 \times 10^6$ light-years $\approx 6 \times 10^5$ pc

Distance of farthest galaxy seen $\sim 3 \times 10^9$ light-years $\approx 10^9$ pc
optically

and in this preliminary survey of cosmic distances we instead make use of
light time. The speed of light *in vacuo* is

▶ $\qquad c = 2.998 \times 10^8$ m s^{-1}. $\qquad\qquad\qquad\qquad\qquad\qquad$ (2.1.1)

Light travels seven times round the Earth in one second, so that the circumference of the Earth is about 'one-seventh of a light-second'. Table 1
shows cosmically important distances measured in light time (the final
column gives the distance in a different unit, the *parsec* (pc), which we
shall introduce in section 2.2.1).

There may well be other matter in the universe besides the galaxies we
see, for instance galaxies that have ceased to radiate, black holes of all
sizes (see section 5.6), and intergalactic dust and gas, but firm experimental evidence for this is lacking. However, a whole range of exotic
astronomical objects has been discovered, most puzzling being the quasi-
stellar objects (QSO's); these appear to be as distant and as bright (both
optically and at radio frequencies) as galaxies, but they are very compact –
at most a few thousand light-years across. Their nature is unknown.
Ignoring these complications, the *mass density* ρ_{gal} due to galaxies has been
estimated to be

▶ $\qquad \rho_{gal} \approx 3 \times 10^{-28}$ kg m^{-3}. $\qquad\qquad\qquad\qquad\qquad\qquad$ (2.1.2)

The universe also contains *radiation*, apparently at all frequencies. Some
of this radiation is *directional*: it comes from localized objects, and indeed
that is how we see them. This leads us to the simplest cosmological observation: the sky is dark at night. The significance of this has been stressed by
Bondi: it tells us that the universe cannot consist of a static, infinitely-old,
infinitely-extended arrangement of galaxies, because if it did every line

drawn outwards from the Earth would terminate on the surface of a star, so that the whole sky would blaze as brightly as the Sun. Absorption by interstellar matter would not prevent this, because the matter would eventually reach thermal equilibrium with the radiation and re-emit with the same intensity as it absorbed. Nevertheless, there are a number of ways out of what used to be called 'Olbers' paradox'. For example, the universe may be infinite in extent, but not infinitely old; then the light from distant stars would not yet have reached us. An alternative escape route from the 'paradox' is the 'expansion of the universe', which we shall discuss in section 2.3. (See also section 6.4.)

The universe also contains *isotropic background radiation*, that is, radiation which is not directional. Energetically, the dominant background radiation is in the microwave region (wavelengths of order 1 mm), and does not come from discrete sources. It seems to have a black-body distribution, which means that it is, or was, in thermal equilibrium with matter; the black-body temperature is 2.7 K. The mass equivalent of this radiation is negligible in comparison with the galaxies; in fact its density (from $E = mc^2$) is

$$\rho_{\text{rad}} \approx 10^{-3} \rho_{\text{gal}}. \tag{2.1.3}$$

Orders of magnitude weaker still are X-ray, γ-ray and radio isotropic backgrounds; these are thought to come from discrete sources, very numerous and too distant to be resolved. Thus if mass dominates the dynamics of the universe (as we believe), then it is safe to neglect the radiation at the present time. But we shall see in chapter 8 that there is good reason to believe that the universe was radiation-dominated in the past, and that the observed microwave background is a fossil remnant of those ancient times.

2.2 The cosmic distance hierarchy, and the determination of galactic densities

How do we know the distances and densities just quoted? The Universe is charted by a sequence of techniques, each of which takes us out to a greater range of distances – to the next level of the 'cosmic distance hierarchy'. Each level is less reliable than the last, so that there is considerable uncertainty about measurements of very great distances.

Before going into the details of these methods of measuring cosmic distances, we employ the 'parable of the city' to illustrate the methods involved. A Martian lands in the dead of night on the flat roof of a building in London. 'He' wishes to make a map of the city, but is confined to

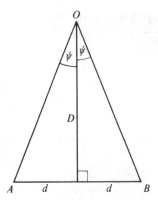

Figure 1. Distance from parallax measurement.

the roof, on which he can see the faint outlines of objects. Outside, only lights are visible – street lights, traffic lights, late-burning room lights, etc. Distances between objects on the roof can be found by direct comparison with any convenient standard, such as the length of the Martian's foot. Distances of lights in the immediate neighbourhood can be found by triangulation using a baseline joining any two points on the roof. But the size of the roof is limited, so that to map more distant lights he must devise another method. In fact he employs the decrease in their apparent *luminosity* with distance: the assumption is made that lights with the same spectral characteristics (such as the red–orange–green of traffic lights, or the yellow of sodium lamps) are physically identical, so that the inverse-square law can be used to determine distances. Provided there are 'standard candles' available for calibration in the triangulable 'near zone', this method works out to distances at which the lights are so faint as to be barely visible. Farther away, the only visible objects are accumulations of lights – apartment blocks, outlying villages – whose distances can again be determined from luminosity measurements, provided the nearest such objects contain visible traffic lights, etc., whose distances are known. Thus the Martian maps the city by using a hierarchy of methods of overlapping applicability, and thus transcends the limitations imposed by his restricted vantage-point.

2.2.1 *Parallax*

This is triangulation using a baseline *AB* whose length 2*d* is known (figure 1), oriented so that the perpendicular from *O*, the object whose distance *D* is to be measured, joins *AB* at its midpoint. As we move from *A* to *B*, the direction of *O* will change. This can be observed as an apparent displacement of O relative to objects far more distant (whose direction hardly alters). If the angular displacement of O is 2ψ, then D follows by elementary trigonometry as

$$D = d \cot \psi \approx d/\psi$$

since in practical cases ψ is a very small angle. The angle ψ (not 2ψ) is called the *parallax* of O relative to AB.

In this way distances within the solar system can be measured, using baselines on the Earth, whose lengths can be established by direct measurement with flexible tapes, or by laser or radar ranging (measurement of the time delay between the emission of a pulse from A and its reception after reflexion from B), or by local triangulation using smaller baselines. Hipparchus measured the Moon's distance by parallax in 129 B.C., and in 1672 Cassini and Richer measured the distance to Mars, using a baseline of about 10^4 km between Paris and Cayenne in South America. This measurement enabled the distance $r_\odot$ from the Earth to the Sun to be calculated, using gravitational theory and the periods $T_\oplus$ and $T_{\mars}$ of orbital revolution of the Earth and Mars. Nowadays distances within the solar system can be measured more accurately using radar.

The mean Earth–Sun distance $r_\odot$ is called the *astronomical unit* (a.u.); it provides the scale for the fundamental baseline from which we can step outside the solar system. This baseline is the diameter of the Earth's orbit, and enables the distances of nearer stars to be determined by measuring their parallax after six months, using the distant stars (whose parallax is negligible) as a background. The basic cosmic distance unit is defined in terms of parallax: one *parsec* ('parallax-second') is the distance of an object whose parallax ψ is one second of arc. Thus

$$1 \text{ parsec (pc)} = \frac{1 \text{ a.u}}{1'' \text{ in radians}} = \frac{3600 \times 180}{\pi} \text{ a.u.}$$

$$= 206\,265 \text{ a.u.}$$

$$= 3.086 \times 10^{16} \text{ m}$$

$$= 3.26 \text{ light-years.}$$

It was in 1837 that Bessel measured the first stellar parallax, of $0.3''$, for the star 61 Cygni (for a brief description of astronomical nomenclature, see Appendix A); the distance of 61 Cygni from us is $1/0.3 \sim 3.3$ pc. The parallax of the nearest star is $0.8''$. The smallest parallax that can be

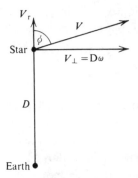

Figure 2. Distance from velocity measurement.

measured is determined by the resolving power of the largest telescopes, and this limits us to stars closer than about 30 pc. A sphere of this radius includes many thousands of stars.

2.2.2 *Distance from velocity measurements*

The 'fixed stars' in fact move, with velocities **V** which can conveniently be resolved into two components: a radial velocity V_r along the line of sight (figure 2) and a transverse velocity $V_\perp$ perpendicular to the line of sight. The magnitude of **V** rarely exceeds 100 km s^{-1}. We can measure V_r by the Doppler shift $\Delta\lambda$ of a spectral line of 'rest wavelength' λ in the light from the star. λ is identified by comparison with patterns of lines observed in terrestrial laboratories. Then, in the non-relativistic limit,

▶ $V_r = c\Delta\lambda/\lambda.$ (2.2.1)

A positive V_r (recession) is indicated by a shift of the lines towards the red end of the spectrum, and a negative V_r (approach) is indicated by a blue shift. It is not possible to measure $V_\perp$ directly, but for nearer stars we can observe the *angular velocity* ω which is due to $V_\perp$; in fact

▶ $D\omega = V_\perp,$ (2.2.2)

where D is the unknown distance of the star. ω is called the *proper motion* of the star; over thousands of years these motions lead to changes in the appearance of the constellations.

D cannot be determined simply by measuring V_r and ω, but there are two limiting cases where extra information is available. In the first, we use clusters of stars which all have approximately the same velocity **V**. This parallel motion can be recognised by the convergence of the proper motions on a point (direction) in the sky. The angle ϕ between this direction and the line of sight to the cluster is the angle between V_r and $\mathbf{V}$, and we have

$$\tan \phi = V_\perp/V_r = D\omega/V_r.$$

Therefore

▶ $\qquad D = V_r \tan \phi/\omega,$ $\qquad\qquad\qquad\qquad\qquad$ (2.2.3)

involving three measurable quantities. This *moving cluster method* works best for open clusters; it has been applied to only a few cases, including the Hyades (in Taurus) whose 100 stars have been found to lie at a mean distance of 40.8 pc from us.

The second limiting case employs groups of stars with no overall motion, where the stars appear to have random velocities like the molecules in a gas. Then it is assumed that the random motion is 'isotropic', that is, that $(2V_r^2)_{\mathrm{av}} = (V_\perp^2)_{\mathrm{av}}$, otherwise we would be in a privileged position relative to the group, which is unlikely. Thus

$$(2V_r^2)_{\mathrm{av}} = (V_\perp^2)_{\mathrm{av}} = D^2(\omega^2)_{\mathrm{av}},$$

and

▶ $\qquad D = \sqrt{[(2V_r^2)_{\mathrm{av}}/(\omega^2)_{\mathrm{av}}]},$ $\qquad\qquad\qquad\qquad$ (2.2.4)

involving averages over the measured V_r and ω for a sample of stars in the cluster. This *method of statistical parallaxes* takes us out to several hundred parsecs (still well within our galaxy).

Real clusters generally lie between these extremes; there is some random motion, and the centre of mass of the cluster has some motion relative to us.

2.2.3 *Distance from apparent luminosity*

Suppose we know the *absolute luminosity L* of a star or galaxy; this is defined as the total power radiated, in watts. Suppose also that we measure the *apparent luminosity l*; this is defined as the power travelling across unit area at the observer, normal to the line of sight. Then, if there has been no absorption in space we may use the conservation of energy to derive the inverse-square law for l in terms of L and the distance D:

$$L = l \times 4\pi D^2,$$

therefore

▶ $\qquad D = \sqrt{(L/4\pi l)}.$ $\qquad\qquad\qquad\qquad\qquad\qquad$ (2.2.5)

Thus D can be found from measurements of l, provided L is known; the trouble is that generally it is not. However, there are certain classes of objects for which we do know L, usually not very accurately, and all higher levels of the cosmic distance hierarchy are based on these 'standard candles'.

To introduce the method in its simplest form, we repeat an argument used by Newton to estimate the distance of the nearest stars. He assumed that all stars are as bright as the Sun, so that the nearest are the brightest. These are about 10^{11} times fainter than the Sun. Therefore

$$D_{star}/r_\odot = (l_\odot/l_{star})^{\frac{1}{2}} = (10^{11})^{\frac{1}{2}}$$

and

$$D_{star} \approx 3 \times 10^5 \text{ a.u.} \approx 1.5 \text{ pc} \approx 5 \text{ light-years.}$$

By what is really a coincidence, based on the fact that the nearest stars happen to be similar to the Sun, this is nearly right (although Newton himself made a numerical error of a factor of 100).

To proceed further, it is necessary to explain that astronomers do not use the power l as a measure of apparent brightness. Rather, they use a logarithmic measure, the *apparent magnitude m*. This is greater, the fainter the object, and is defined so that two objects whose luminosities l_1 and l_2 differ by a factor of 100 differ in apparent magnitude by 5, that is,

▶ $$l_1/l_2 = 100^{(m_2-m_1)/5}; \quad m_2 - m_1 = 2.5 \log_{10}(l_1/l_2). \tag{2.2.6}$$

We still have to fix the zero point of the magnitude scale. Historically, $m \sim 1$ corresponded to the brightest few visible stars, while $m = 6$ stars are those just visible with the naked eye. Nowadays it is necessary to be more precise, and the zero point is defined by

▶ $$l_{(m=0)} \equiv 2.52 \times 10^{-8} \text{ W m}^{-2} \tag{2.2.7}$$

(with this definition, l is the *total* power received, integrated over all wavelengths). The apparent magnitude of the Sun is $m_\odot = -26.85$.

The *absolute magnitude M* of an object is the apparent magnitude that the object would have at a distance of 10 pc. Thus, measuring D in pc,

$$l_m \times 4\pi D^2 = l_M \times 4\pi 10^2.$$

Therefore

$$100^{(M-m)/5} = 10^2/D^2,$$

▶ $$D = 10^{[1+(m-M)/5]}$$

▶ $$m - M = 5 \log_{10}(D/10). \tag{2.2.8}$$

The Sun has an absolute magnitude $M_\odot = 4.72$, fairly typical for a star. Bright galaxies have $M_{gal} \sim -22$, so that the absolute luminosities $L_\odot$ and

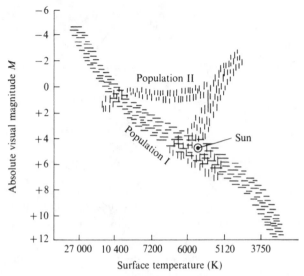

Figure 3. Hertzsprung–Russell diagram (schematic).

L_{gal} (which are proportional to the apparent luminosities at 10 pc) are related by

$$L_{gal}/L_{\odot} = 100^{(M^{\odot}-M_{gal})/5} = 100^{(4.72+22)/5} \sim 5 \times 10^{10},$$

a result consistent with our earlier assertion that large galaxies contain about 10^{11} stars. The quantity $m - M$ is called the *distance modulus* (for the Sun, its value is -31.57).

M, m, L and l have been defined in terms of the total power emitted or received at all wavelengths. However, analogous quantities can be defined in terms of the power per unit wavelength or frequency interval.

We have some knowledge of M for several classes of object, and we shall discuss just four: main-sequence stars, Cepheid variables, novae, and brightest galaxies in clusters.

Main-sequence stars. For nearby stars whose distances can be found by parallax or velocity measurements, Hertzsprung and Russell found in 1910 that for many stars (the 'main sequence') L (or M) and spectral type (roughly the colour, which corresponds to the surface temperature T) are strongly correlated (figure 3); the cooler stars are fainter. Thus if we know that a given star is in the main sequence, we simply measure its apparent magnitude m and determine its spectral type; the latter gives us the absolute magnitude M, from the Hertzsprung–Russell relation, and from the distance modulus $m - M$ we infer D. The method works best for clusters,

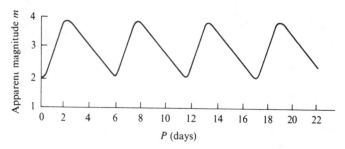

Figure 4. Light curve of δ Cephei.

where all stars are roughly at the same distance, so that the main sequence can be identified in a statistical way, as follows: the *apparent* magnitudes m are plotted against spectral type, and it is often found that the points lie near a curve similar to that resulting from the Herzsprung–Russell relation; any star on the curve is therefore a member of the main sequence. Complications arise, however, because there are (roughly speaking) two main sequences: population I stars (like the Sun) in open clusters like Hyades, and population II stars in globular clusters like M13 in Hercules; great care must be taken to avoid confusing these populations. Main-sequence stars are rather faint (at least where the colour–luminosity relation is reliable) and this limits the distances which can be determined. The largest telescope cannot detect stars fainter than about $m = 22.7$. If such a star is seen to have the same colour as the Sun ($M = 4.7$) then the distance modulus is 18, and the actual distance is

$$D = 10^{1+18/5} \text{ pc} = 10^{4.6} \text{ pc} \approx 4 \times 10^4 \text{ pc}.$$

This is comparable with the diameter of our galaxy, so that even this giant step up to a new level in the hierarchy has not taken us out to cosmologically interesting distances.

Cepheid variables. Many stars vary regularly in brightness, with periods P of the order of days. Typical is δ Cephei, whose 'light curve' is shown in figure 4. These variable stars are therefore called 'Cepheids' (they should not be confused with pulsars, which are radio sources and whose periods are much shorter). It was observed by Leavitt in 1912 that m and P are approximately linearly related for the Cepheids in the Small Magellanic Cloud (now known to be a 'satellite' sub-galaxy of our own). Since these stars are all roughly at the same distance from us, she concluded that m was uniquely related to M, and hence M to P, thus giving an absolute luminosity *versus* period relation. Because all Cepheids with the same

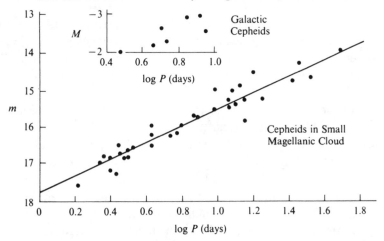

Figure 5. Cepheids as standard candles. (By kind permission of Professor G. C. McVittie.)

period have the same M, these stars may be employed as 'standard candles' for distance determination, once the relation has been calibrated by establishing M for Cepheids within the galaxy, using a lower level of the distance hierarchy. Unfortunately, there are only a few galactic Cepheids in clusters whose distance is known, and this reduces the accuracy of the method. Figure 5 shows some actual data.

Thus a Cepheid with period 10 days ($\log P = 1$) has an absolute magnitude $M \approx -3$ (i.e. $L/L_\odot \sim 10^3$). In the Magellanic Cloud a physically similar object (same M) has an apparent magnitude m of about 16. Therefore the distance modulus $m - M$ is about 19, and the distance is

$$D = 10^{1+19/5} \text{ pc} = 10^{24/5} \text{ pc} \approx 6 \times 10^4 \text{ pc.}$$

At last we are outside the galaxy, but only just!

Now Cepheids are intrinsically rather bright (up to $M \sim -6$, i.e. $L/L_\odot \sim 10^4$), and they can be resolved in a number of external galaxies as well as our own. In 1923 Hubble discovered Cepheids in M31 (the Andromeda nebula) and measured their periods. Their faintness proved conclusively that 'spiral nebulae' are galaxies outside our own but similar to it – i.e. 'island universes' – and thus settled a long controversy, establishing our presently-held picture of the universe as an assembly of galaxies. However, his value of the distance D was 2.8×10^5 pc, and this was too small by a factor of more than two, as was shown by Baade in 1952, because Hubble had confused two different types of variable star. Hubble's erroneous value led to the puzzling conclusion that all other galaxies are much smaller than our own, which would give us a privileged status as observers; but the new value gives our galaxy an unembarrassingly typical size.

Cepheids can be seen out to about 4×10^6 pc, which includes the galaxies of the 'local group' – the cluster of galaxies containing our own.

Novae. In each large galaxy close enough to be studied in detail, about forty stars are observed each year to flare up suddenly and become up to six magnitudes brighter than they were before. Then they gradually fade, over a period of days. These are novae, or 'new stars'. They can also be observed in our own galaxy, and their distances determined by methods lower down the hierarchy. Thus the absolute magnitudes M are known. These vary with time, of course, and it is found that the maximum brightness $M_{\max}$ is fairly well correlated with the fading time t_2 (conventionally defined as the time for the star to grow two magnitudes dimmer, that is from $M_{\max}$ to $M_{\max} + 2$). Thus novae can be used like Cepheids, to infer D from measurements of t_2 and the apparent maximum brightness $m_{\max}$. The results of the method for neighbouring galaxies agree reasonably well (within 10 per cent) with distances deduced from Cepheid observations. However, novae are much brighter than Cepheids, so they constitute another level of the hierarchy. How far does this take us? The brightest novae reach $M_{\max} \sim -9.3$, and we can currently detect $m_{\max} \sim 22.7$, so that the limiting distance with this method is

$$D = 10^{1+[22.6-(-9.3)]/5} \text{ pc} = 10^{38/5} \text{ pc} = 4 \times 10^7 \text{ pc}.$$

We have now reached out beyond the nearest galaxies outside our local group; these lie in the Virgo cluster, which contains about 2500 galaxies. Extra checks on the estimation of distances on this level of the hierarchy are obtained by using globular clusters of stars, and HII regions (clouds of ionised hydrogen surrounding hot stars) as standard candles.

Brightest galaxies in clusters. Beyond the Virgo cluster it is not easy to resolve individual stars, and the final level of the distance hierarchy employs whole galaxies as distance indicators. It appears that the distribution of apparent galactic brightnesses within a cluster has a rather sharp upper limit. For the Virgo cluster we know the distance, and hence the absolute magnitude M of the brightest galaxy; its value is -21.7. If we assume the brightest galaxy in a distant cluster and has the same M, then we can find D simply by measuring the apparent magnitude m. This takes us out to

$$D = 10^{1+[22.7-(-21.7)]/5} \text{ pc} \approx 8 \times 10^9 \text{ pc},$$

which at last includes a large fraction of cosmologically interesting objects.

But there is a snag: suppose that the distribution of galactic brightnesses has a small 'tail' instead of a sharp cut-off, that is, that there is a small fraction of extraordinarily bright galaxies. Then as we probe to greater distances, and choose increasingly rich clusters, we shall inevitably encounter these exceptional galaxies whose absolute brightness exceeds $M = -21.7$, and thus underestimate all their distances. This is the 'Scott effect', pointed out in 1957; its existence is still controversial. A more detailed discussion of the cosmic distance hierarchy (with references) may be found in Weinberg's *Gravitation and cosmology* (see the bibliography).

At this point it is convenient to draw attention to two conceptual problems that will occupy us a great deal later on. The first problem concerns the simple distance formulae given by the parallax, velocity and luminosity methods. The formulae involve elementary trigonometric relations between angles and distances for systems of straight lines. These are based on ordinary *Euclidean geometry*, together with the identification of light rays with straight lines. This is known to apply very accurately in the solar system, because the various very precise methods for determining distance agree with one another. But can we extrapolate to galactic or intergalactic distances? There are, after all, various systems of non-Euclidean geometry, such as that which must be used on the surface of a sphere (in this geometry angles of a triangle do not add up to 180°, circles of radius r have circumferences less than $2\pi r$, etc.). It is a question not of mathematics but of physics to ask which geometry applies in the actual world. We shall find strong evidence that 'space may be curved', and this will require us to talk very carefully about distance. However, within the 'local group' of galaxies and even a little beyond (i.e. out to many megaparsecs) any corrections for possible non-Euclidicity are negligible.

The second conceptual problem concerns time. The cosmic hierarchy takes us out to billions of light-years. Thus we are seeing light emitted in the remote past. What is the universe like now? This question assumes that distant events can uniquely be considered simultaneous, and it is known from the special theory of relativity that events simultaneous in one frame of reference will not be so in others: there is no unique 'now'. Thus we must take great care in our description of the Universe on a large scale, particularly since we believe it to be evolving as time proceeds.

2.2.4 *Weighing galaxies*

In the models of the universe we shall derive from general relativity, the mass density ρ is a vitally important quantity. At present the mass seems to occur principally in the form of galaxies, and we have quoted a value ρ_{gal} of 3×10^{-28} kg m^{-3}; how has this figure been arrived at? By employing the distance hierarchy and counting galaxies, the number density (galaxies per unit volume) can be found. Thus it is necessary to find the average mass of a single galaxy. There are several ways of doing this, all based on the (far from obvious) assumption that galaxies, or clusters of galaxies, are *gravitationally-bound systems in dynamical equilibrium*. The parts of such a gravitationally-bound system must be in relative motion, otherwise its contents would fall in towards the centre of mass. Indeed galaxies are observed to rotate: points equidistant from the axis of galaxies seen almost edge-on have different spectral shifts, that is, different Doppler velocities. The average of these gives the mean velocity of the galaxy as a whole, while half their difference gives the rotational velocity $V(r)$ about the centre, which varies with radial distance r.

The mass M_g of a galaxy can be estimated as follows: this mass, acting as though it were concentrated at the galactic centre, determines the motion of the outermost parts of the galaxy, whose radius is r_g, say. (Strictly, this would apply only to a spherically-symmetrical galaxy, but we are only making rough estimates here.) If galaxies are stable, the orbits of stars in them will approximate to closed curves. If we assume these are circles, then, for a star of mass m on the edge, we have, from Newton's law:

$$\text{Mass} \times \text{acceleration inwards} = mV_g^2/r_g = \text{inward force}$$
$$= mM_g G/r_g^2,$$

that is

▶ $$M_g = r_g V_g^2/G, \qquad (2.2.9)$$

where G is Newton's gravitation constant and V_g stands for $V(r_g)$. Both r_g and V_g are measurable, so M_g can be found. The equation applies also (and more strictly!) to the motion of the Earth round the Sun, so that (2.2.9) can be written alternatively as

▶ $$M_g/M_\odot = r_g V_g^2/r_\odot V_\oplus^2, \qquad (2.2.10)$$

where $r_\odot$ is the Earth–Sun distance ($\sim 5 \times 10^{-6}$ pc) and $V_\oplus$ the Earth's orbital speed (~ 30 km s^{-1}). (Note that the symbol M here stands for mass, not magnitude as before.)

Let us apply this to our own galaxy: we have $r_g \sim 2 \times 10^4$ pc, and the Sun's orbital speed about the galactic centre is about 200 km s^{-1} (this is obtained by a statistical analysis of the Doppler shifts and proper motions of a large number of stars). The Sun is about halfway out, so we take $V_g \sim 200/\sqrt{2}$ km s^{-1} (this assumes the galaxy to be fairly concentrated near its centre, which even casual observation of the Milky Way shows to be the case). Then

$$M_g/M_\odot = 2 \times 10^4 \times (200)^2/5 \times 10^{-6} \times 2 \times (30)^2 = 9 \times 10^{10}.$$

More sophisticated analyses all give around $10^{11} M_\odot$ for M_g, and this value seems typical for large galaxies, Note that because we have used *dynamical* arguments, rather than, say, star counts, our value for M_g includes invisible mass in the form of interstellar dust, black holes, dead stars, etc. There may of course be mass *between* the galaxies (dust, hydrogen, etc.), and this 'missing mass problem' is a question of intense current interest. Dynamical analysis of whole clusters of galaxies at one time suggested that there might be up to a hundred times more mass between galaxies than in them, but there is growing doubt about this.

To find the total density ρ_{gal} it is necessary to know the number density n of galaxies in space. This is found as follows. Simplify the problem by assuming that the galaxies all have the same absolute luminosity L. Then all galaxies whose *apparent* luminosity exceeds a given value l must lie within the sphere of radius D_l, where

▶ $$D_l = (L/4\pi l)^{\frac{1}{2}}.$$ (2.2.11)

The total number $N(> l)$, of such galaxies can be observed, and is given by

$$N(> l) = n \times \text{volume of sphere of radius } D_l = \tfrac{4}{3}\pi n D_l^3$$

or

▶ $$N(> l) = \tfrac{4}{3}\pi n (L/4\pi l)^{\frac{3}{2}} = \tfrac{4}{3}\pi n \times 10^{[3+3(m-M)/5]}.$$ (2.2.12)

where n is in pc^{-3}. The $l^{-\frac{3}{2}}$ dependence is quite well obeyed for galaxies fairly close to us, so that n can be determined; its value is about one 'average galaxy' per 75 cubic megaparsecs; thus, roughly,

$$\rho_{gal} \sim 10^{11} M_\odot \, pc^{-3}/75 \times 10^{18} \sim 10^{-28} \text{ kg m}^{-3},$$

whereas the current 'best value', based on elaborate analysis taking account of the fact that the galaxies do not all have the same luminosity and mass, is about three times greater. (It is worth mentioning at this point that the $l^{-\frac{3}{2}}$ law is not universally valid: it breaks down for distant radio sources, in a way which sheds considerable light on the history of the universe, as we shall see in chapter 6. For faint *stars* the law does not hold at all, and this is another indication that the galaxy is finite.)

2.3 The red shift and the expansion of the universe

Between 1910 and 1930 the cosmic distance hierarchy reached out beyond 10^4 pc, and it was realised that the universe consisted of a vast number of other galaxies resembling our own. It was natural at the same time to investigate the motion of these galaxies, by looking for Doppler shifts in their spectra, in order to reveal motion along the line of sight. Let λ_e be the wavelength of a spectral line in the rest frame of its source, and let λ_o be the wavelength measured by an observer moving relatively to the source with velocity v. Then the *spectral shift z* is defined as

▶ $\qquad z \equiv (\lambda_o - \lambda_e)/\lambda_e.$ $\qquad\qquad\qquad\qquad$ (2.3.1)

($z > 0$ means a red shift, $z < 0$ means a blue shift). Interpreting z as a Doppler shift and using the familiar formula implies – at least for small z – that

▶ $\qquad v = cz$ $\qquad\qquad\qquad\qquad\qquad\qquad\qquad$ (2.3.2)

($v > 0$ means recession from us, $v < 0$ means approach to us).

It was expected that z would vary from galaxy to galaxy in a random manner, being as often negative as positive; indeed this is roughly true for the local group. Beyond the local group, however, z is always positive, indicating that *all these other galaxies are receding from us*. These motions, originally discovered in 1915 by Slipher, were studied very carefully by Hubble, who announced in 1929 the famous result that the speed of recession of a galaxy is proportional to its distance from us: ·

$$v = HD,$$
$$H = (55 \pm 7) \text{ km s}^{-1} \text{ Mpc}^{-1} = (1 \pm 0.1)/(1.8 \times 10^{10}) \text{ year}^{-1}.$$

$\qquad\qquad\qquad\qquad\qquad\qquad\qquad\qquad\qquad\qquad\qquad$ (2.3.3)

The formula for v is *Hubble's law*, and H is Hubble's constant. The value quoted is widely accepted, but some astronomers consider lower values quite possible. The considerable uncertainty in H arises from the unreliability of the later stages of the cosmic distance hierarchy and not from scatter in the basic data of z v. m for the brightest galaxies in clusters, as figure 6 shows. H is a crucial parameter in all cosmological theories, so much current effort is being put into measuring it more precisely, by a variety of methods.

What does it mean to have all the galaxies receding from us? Expressed vectorially, Hubble's law asserts that the velocity **v** of a galaxy G at a position **r** measured from our own galaxy as origin (figure 7) is

$$\mathbf{v} = H\mathbf{r}.$$

Now consider G from the point of view of a third galaxy G' at **r**′ relative to us, and therefore moving away from us with velocity $\mathbf{v}' = H\mathbf{r}'$. The velocity of G relative to G' is

$$\mathbf{v} - \mathbf{v}' = H\mathbf{r} - H\mathbf{r}' = H(\mathbf{r} - \mathbf{r}'),$$

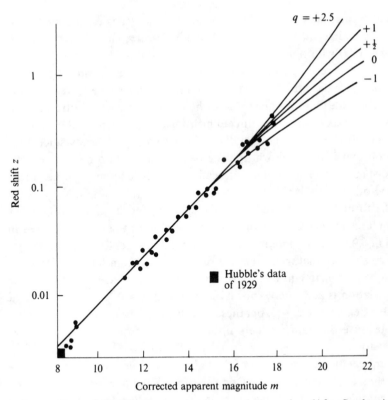

Figure 6. Hubble diagram showing cosmic expansion. (After Sandage.)

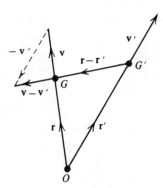

Figure 7. Expansion as seen by O and G'.

so that G' also sees G, and therefore all the other galaxies, receding from itself. Of course, we have used Euclidean geometry, and ignored possible changes in H with time, but the general result still holds: each galaxy sees all the others receding from itself. Now, on large scales (greater than about 40 Mpc), galaxies are observed to be uniformly distributed over the sky, so we conclude that the universe as a whole is homogeneous, isotropic, and, above all, *expanding*. This three-dimensional expansion of the universe is often illustrated by the two-dimensional analogy of beads stuck to a balloon that is being blown up. The beads (galaxies) do not expand but the distance between any pair of beads increases, and any one bead sees all the others receding from itself.

The formula $v = cz$ holds only for small z; for larger z, the Doppler formula from general relativity must be employed; this will be derived in section 6.2. A similar modification (section 6.3) applies to the $z–m$ relation, and we shall see that measurements of departures from linearity for large z can help to distinguish various possible model universes. At present the $z–m$ relation is accurately known out to about $z \sim 0.1$. The red shifts of QSOs extend out to $z \sim 3$, but these objects cannot be used to test Hubble's law because they differ widely in absolute luminosity in a way not yet understood.

Our belief in the expansion of the universe rests principally on the interpretation of the red shift as a Doppler effect. What alternatives are there? We consider just two. First, there is the *gravitational red shift*; this is a consequence of general relativity (or indeed, any of a wide class of gravitational theories) which can, however, be understood in a simple way: to escape from a gravitational field a photon must expend energy. After escape it has lost this energy and so is redder. Consider a photon emitted with frequency ν_e and escaping to infinity from the surface of a galaxy of radius r_g and mass M_g. If the frequency observed at infinity is ν_0, we have

$$\text{Energy loss} = h(\nu_e - \nu_0) = \text{work done} = \frac{GM_g}{r_g} \times (\text{'mass' of photon})$$

$$= \frac{GM_g}{r_g} \frac{h\nu_0}{c^2}.$$

But $\nu = c/\lambda$, so that the red shift defined by (2.3.1) is

$$z \equiv \frac{\lambda_0 - \lambda_e}{\lambda_e} = \frac{\nu_e - \nu_0}{\nu_0} = \frac{GM_g}{r_g c^2}. \tag{2.3.4}$$

Later we shall see how this relation is verified by precise experiments near the Earth and Sun. However, if we put in the numbers for a typical galaxy, we find $z \sim 10^{-7}$. This is negligible in comparison with observed galactic shifts, and in any case the model would not explain the distance-dependence in Hubble's law. Therefore we can confidently assert that the red shifts are not gravitational in origin.

The second possible interpretation is the *tired light effect*: light travelling over vast distances would lose energy and therefore redden. The energy loss could not arise by the inelastic scattering of photons by the particles of some intergalactic medium, because (*a*) such scattering would produce a distribution of energy loss, and not the single red shift observed, and (*b*) inelastic scattering would deflect the photons, thus blurring out the images of distant galaxies, and this is not observed. Therefore the tired light effect is sometimes postulated as a new property of electromagnetic radiation. This would certainly explain the red shift without motion of the galaxies, but there is a deep philosophical objection to explanations of this sort. They are *ad hoc hypotheses*. This means that they are introduced to account for a particular experimental result, and remain disconnected from the rest of physical knowledge. Such explanations are unsatisfactory because they are not open to experimental disproof: they are not *falsifiable*. It is always possible to 'explain' any experiment by postulating a new effect, but the difficulty (and the challenge) in science is to unify extensive bodies of knowledge. This is exactly what the Doppler explanation does: changes in frequency resulting from relative motion of source and observer have been well-verified in the laboratory for all types of wave motion. It is, of course, ultimately a matter of experiment to settle the question of whether or not red shifts are Doppler effects, and each type of object must be tested separately. At present the 'Doppler' interpretation is much more firmly established for galaxies than for quasars. From now on, however, we shall interpret red shift data as indicating that the universe is expanding.

The most dramatic inference from Hubble's law is that in the past the galaxies were much closer together than they are today. In fact at a time of order $H^{-1} \sim 2 \times 10^{10}$ years ago the matter density of the universe may have been very large. This is the 'big bang' theory of the origin of the universe. There are three main independent bodies of evidence in its favour. *Astrophysical theory* can now account fairly well for the existence of the main sequence of stars, and explain the course of stellar evolution in our own and other galaxies in terms of nuclear physics and the transfer of radiation in a gravitational field. All the evidence points to typical stellar ages of 10^{10} years or less. *Radioactive isotopes* are assumed to have been produced in certain abundance ratios by nucleosynthesis, and to have subsequently decayed to their present abundances, which therefore give an indication of the time that has elapsed since their production, because the decay rates are known. The 'ages of the elements' deduced in this way are about 10^{10} years. Being less than 2×10^{10} years, both values are consistent with a 'big bang' origin for the universe. *The microwave radiation background* strongly suggests a dense, radiation-dominated early phase of the universe, as we shall see in chapter 8.

However, there is an important alternative to this evolutionary picture of the universe, namely the *steady state theory*. This is based on an extension of the cosmological principle to include time as well as space. The new 'perfect cosmological principle' states that the large-scale universe not only looks the same to every observer, but also that it always has looked the same, and always will do so. To keep the density constant even though the universe is expanding requires the 'continuous creation' of matter from nothing, at the unobservably small rate of about one hydrogen atom per 6 km^3 per year. Metaphysically the theory has attractive features: at the cost of giving up energy conservation the problem of 'what happened before the big bang' is avoided. Further, when put in relativistic form the steady state model gives very definite predictions for the way in which Hubble's law should cease to hold for large z, and for the law which should be obeyed by galactic number counts as a function of apparent magnitude m. Therefore the model is *falsifiable*, and not at all *ad hoc*. The trouble is that experiment does seem to contradict it, and there is, moreover, no obvious explanation for the microwave background. The details of the steady state model will be explored in section 7.3.

Another inference from Hubble's law, if we assume that Euclidean geometry may be employed, is that the galaxies at a distance

$$D_{\max} = c/H \sim 2 \times 10^{10} \text{ light years} \sim 6 \times 10^9 \text{ pc}$$

are receding as fast as light. Light from more distant galaxies can never reach us, so that $D_{\max}$ marks the limit of the observable universe; it corresponds to a sort of *horizon*. We shall discuss horizons more fully in section 6.2, when we have learned some general relativity; there we shall learn that the 'visibility-in-principle' of distant galaxies is strongly affected by departures from Euclidean geometry.

3 *Physical basis of general relativity*

3.1 The need for relativistic ideas and a theory of gravitation

The emission of light from a distant galaxy is an *event*; our reception of this light is another event. To discuss events in physics, that is, to correlate and explain them, we need a *frame of reference*, within which events are located with the aid of a *coordinate system*. Each event is specified by four numbers, usually taken as three position coordinates **r** and one time coordinate t. The simplest frame of reference for the location of cosmological events, such as the reception of light from a galaxy, is a frame at rest relative to our own galaxy, or to the centre of mass of the local group of galaxies. (The frame of our earthly observatories is a reasonable approximation to this for most purposes, because its speed relative to local galaxies is about 200 km s^{-1}, and for galaxies more distant than 10 Mpc this is swamped by the general recession.)

However, the simplest frame for locating an event such as the emission of light from a distant galaxy G is not our local frame, but the local frame of G. In this frame, let the emission event have coordinates $\mathbf{r}_G$, t_G. What are the coordinates **r**, t of this event in our own frame? This is a question of *relativity*: how can we *transform* the coordinates of an event in one frame, to get the coordinates in another? Within *Newtonian mechanics*, transformation between frames involves only the position coordinates **r** and $\mathbf{r}_G$, and the relative velocity and acceleration. The time coordinates do not transform, since it is assumed that $t = t_G$. In Newton's words: 'Absolute true and mathematical time, of itself, and from its own nature, flows equably without relation to anything external.' In the special case where the two frames are inertial (see section 3.3), the Galilean transformation (defined by (3.4.1) later) must be used in Newtonian mechanics. But we know now that Newtonian mechanics is only an approximation, valid for speeds small compared with c; between two inertial frames moving with arbitrary relative speed, Einstein's *special theory of relativity* tells us that the Lorentz transformation (see (3.4.2) later) must be employed to

[23]

relate the two sets of coordinates. This involves abandoning the concept of absolute time, and leads to the well-known 'paradoxes' – effects concerning the different rates of relatively moving clocks, which have now been well verified experimentally. An excellent treatment of special relativity forms the first part of Rindler's *Essential relativity* (see the bibliography).

Now consider again our local frame and the local frame of G. If the distance of G is comparable with H^{-1} light-years, then its recession velocity is comparable with c, and the transformations of Newtonian mechanics cannot be employed. Unfortunately, however, the Lorentz transformation cannot be used instead, because our galaxy and G may be mutually *accelerating* under gravity, a situation to which, as we shall see, special relativity cannot validly be applied.

Therefore, in order even to *describe* events properly on a cosmic scale, we need a further extension of Newtonian mechanics, beyond special relativity; we need a *general theory of relativity*, to discuss the transmission of light signals and the relationship between the coordinates of frames of reference which are arbitrarily accelerated relative to one another.

However, it is clear that a description of the universe – even a properly relativistic one – is not enough. We require a physical theory which *explains* the configuration and motion of the system of galaxies, that is, we want a dynamics as well as a kinematics, to make a complete *mechanics*. At the outset we can neglect all quantum effects, because these are only important for small masses and on a microscopic scale. Mechanics is the study of the motion of a system as it results from interactions between its parts and influences from outside. In Newtonian mechanics, 'interactions' and 'influences' are described by *forces*. What forces act on a galaxy? Clearly none from 'outside' the system of other galaxies, since we know of no constituents of the universe at large other than galaxies. How then do galaxies interact with one another? Through the action of *gravitation*; all other forces are negligible.

This is not quite as obvious as it seems, because by any reasonable measure gravity is the weakest force in physics. To compare with *electrostatic forces*, for example, we can calculate the ratio of gravitational to electrostatic forces between an electron and a proton; this is the same as the ratio of energies required to break the gravitational and electrostatic bonds between the two particles. The result of the calculation is 4×10^{-40}, which is why we neglect gravitation in atomic physics. For two protons inside a nucleus, the ratio is 10^{-36}, so that gravitation is also negligible in nuclear physics. What about *nuclear forces*? Their strength can be estimated from the work needed to pull a nucleon out of a nucleus; using data for the binding energy of α particles, we can show that the nuclear force between two protons in a nucleus is about ten times stronger than the electromagnetic force, and therefore 10^{37} times stronger than gravity. Thus on the very smallest scales – within atoms and nuclei – gravity is utterly negligible. This is also true on the next scale: interactions *between* atoms, which give rise to the cohesion of matter in bulk. The work required to separate two atoms or molecules (that is, the strength of the 'atomic bond' in a solid) is roughly kT_m where T_m is the melting temperature and k is Boltzmann's constant. For sulphur,

$$kT_m \sim 5 \times 10^{-21} \text{ J.}$$

By contrast, the work required to separate two sulphur atoms 2 Å apart if gravity were the only force between them would be

$$\frac{G \times (\text{mass of sulphur atom})^2}{2\text{Å}} \sim 10^{-51} \text{ J,}$$

where G is the gravitational constant. (Before atomic spacings were known, it was thought that matter might be gravitationally bound; this is quite impossible, according to our calculation.)

Weak though it is, gravity has two properties that make it dominate all other forces on the astronomical scale. Firstly, it is a *long-range* force, falling off only as $1/r^2$. By contrast, nuclear forces fall off roughly as $\exp(-\alpha r)/r^2$, and are negligible for nucleons separated by more than about 10^{-14} m; similarly, interatomic forces fall off roughly as $1/r^7$, and are negligible for two atoms separated by more than about 10^{-9} m. Thus these 'local' forces are insignificant in systems composed of bodies widely separated in a vacuum, by comparison with gravitational and electrostatic forces. The second property of gravity is that it is *unscreened*: only positive masses exist. Electric charges can be negative as well as positive and matter as a whole is neutral. Thus the vast electrostatic repulsion exerted on our protons by the protons in the Moon is cancelled – 'screened' – by the almost precisely equal and opposite attraction by the electrons in the Moon. It has occasionally been suggested that matter in bulk may not be quite neutral. Perhaps the electron charge is very slightly different from the proton charge, or perhaps the total numbers of positive and negative charges are not quite the same. Such a net charge on the universe would cause a 'cosmic repulsion' that might explain the observed expansion. To achieve this, however, the repulsion must dominate over gravity, and this is certainly not true locally, that is, in the solar system, so that the charge imbalance would have to be very inhomogeneously distributed. There is no

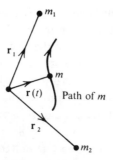

Figure 8. Path of a particle.

experimental evidence at all for this violation of charge neutrality, so, while we cannot rule it out, it remains an *ad hoc* hypothesis, explaining just one fact – expansion – and unconnected with the rest of our knowledge.

Gravity, then, is both long-range and unscreened, and so predominates over cosmic distances. Therefore any cosmology must be founded on a logically secure theory of gravitation. We shall show in the next few sections how Newton's theory, although amazingly accurate for the solar system, astronautics etc., is nevertheless unsatisfactory, and we shall follow the steps which led Einstein to replace it with a new theory which at the same time provided the general theory of relativity between arbitrary reference frames that we have already found to be needed in cosmology.

3.2 Difficulties with Newtonian mechanics: gravity

Gravity is a very strange force in Newtonian mechanics. Consider Newton's second law:

$$\blacktriangleright \qquad m\mathbf{a} = \mathbf{F}. \tag{3.2.1}$$

This tells us that the total force $\mathbf{F}$ on a body determines its acceleration $\mathbf{a}$. $\mathbf{F}$ arises from the action of other matter on the body, and may be gravitational, electrical, frictional, intermolecular (e.g. contact forces) etc. The constant m is the mass of the body; it expresses its *inertia*, that is, its ability to resist being accelerated. The greater m, the smaller is $\mathbf{a}$ for a given $\mathbf{F}$. Generally, $\mathbf{F}$ depends on properties of the body which are independent of its mass (charge, size, magnetic moment, etc.) as well as on properties of the other matter acting on it. In the one case of gravitation, however, m also occurs in the force. Let $\mathbf{F}$ be a gravitational force acting on m, arising from N masses $m_1, m_2, ..., m_N$ at fixed positions $\mathbf{r}_1, \mathbf{r}_2, ..., \mathbf{r}_N$ (figure 8).

Let $\mathbf{r}(t)$ be the changing position of m. Then the inverse-square law gives

$$\blacktriangleright \qquad \mathbf{F} = - \sum_{i=1}^{N} \frac{Gmm_i(\mathbf{r} - \mathbf{r}_i)}{|\mathbf{r} - \mathbf{r}_i|^3}. \qquad (3.2.2)$$

Thus the mass m which governs a body's resistance to forces – the *inertial mass* – is also the mass which determines the magnitude of the gravitational force acting on it – i.e. the *passive gravitational mass*.

This dual role played by mass has an astonishing consequence: inserting (3.2.2) into (3.2.1), we find that m cancels, and we get

$$\blacktriangleright \qquad \mathbf{a} = -G \sum_{i=1}^{N} \frac{m_i(\mathbf{r} - \mathbf{r}_i)}{|\mathbf{r} - \mathbf{r}_i|^3}. \qquad (3.2.3)$$

Thus the acceleration of a particle in any gravitational field is independent of its mass. In other words, if only gravitational forces act, all bodies similarly projected pursue identical trajectories. This was known to Galileo; he believed that very different bodies dropped together from a building would hit the ground at approximately the same time, provided they were heavy enough for air resistance to be negligible (i.e. for gravity to be effectively the only force acting). To make a more precise test, he slowed the motion by using an inclined plane, and found that balls of different masses rolled down in the same time.

Now let us define rather carefully the different sorts of mass. Inertial mass m_I occurs in $\mathbf{F} = m_I \mathbf{a}$. Gravitational mass m_G occurs in the inverse-square law (3.2.2), in two ways: *passively* (m_G^p) in determining the force *on* the body from others, and *actively* (m_G^a) in determining the force *with which* the body attracts others. Galileo's experiment suggests that

$$\blacktriangleright \qquad m_I = m_G^p. \qquad (3.2.4)$$

To test the accuracy with which this holds (that is, the accuracy with which different bodies fall with the same acceleration) a series of increasingly precise experiments has been carried out. Newton made pendula of the same lengths but different compositions, and found that they had the same periods. Eötvös in 1889 showed that m_I/m_G^p was unity to within 10^{-9} for wood and platinum; Roll, Krotkov and Dicke in 1964 showed that aluminium and gold fall towards the Sun with the same acceleration to one part in 10^{11}. It has also been shown that neutrons fall with the same acceleration $\mathbf{g}$ on Earth as ordinary matter, and that the gravitational force on electrons in copper is the same as on free electrons. Thus we are justified in treating the equality of m_G^p and m_I as a law of nature.

By contrast, the equality of active and passive gravitational mass is not

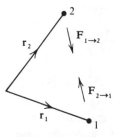

Figure 9. Action and reaction.

an independent law of nature, but follows from Newton's third law. Consider two bodies 1 and 2, at $\mathbf{r}_1$ and $\mathbf{r}_2$ (figure 9), with gravitational masses m_{G1}^a, m_{G1}^p, m_{G2}^a, m_{G2}^p, The force $\mathbf{F}_{2\rightarrow 1}$ that 2 exerts on 1 is

$$\mathbf{F}_{2\rightarrow 1} = -Gm_{G1}^p\, m_{G2}^a\, \frac{(\mathbf{r}_1-\mathbf{r}_2)}{|\mathbf{r}_1-\mathbf{r}_2|^3},$$

while

$$\mathbf{F}_{1\rightarrow 2} = -Gm_{G2}^p m_{G1}^a\, \frac{(\mathbf{r}_2-\mathbf{r}_1)}{|\mathbf{r}_1-\mathbf{r}_2|^3}$$

is the force that 1 exerts on 2. These forces are clearly in opposite directions, as the third law requires, but they must also be equal in magnitude. This leads to

$$m_{G1}^p,\, m_{G2}^a = m_{G2}^p\, m_{G1}^a,$$

that is,

▶ $\qquad m_{G1}^p/m_{G2}^p = m_{G1}^a/m_{G2}^a.$ $\qquad\qquad$ (3.2.5)

Thus active and passive gravitational masses are proportional, and without loss of generality we can define them as equal.

For purely gravitational forces, then, m_I, and m_G^p cancel in Newton's second law. But the active mass m_G^a still plays a role, because the strength of the gravitational field in which all 'test bodies' fall similarly still depends on the value of m_G^a for all the 'source' bodies giving rise to the field. For non-gravitational forces, neither m_G^a nor m_G^p is involved, and motion depends only on m_I. Thus the first difficulty with Newtonian mechanics is: why is gravity such an anomalous force? How can we frame a gravitational theory from which the equal acceleration of all masses in a gravitational field follows naturally, instead of having to be added (in the form (3.2.4)) as a separate law of nature?

The second difficulty with Newtonian mechanics is a stubborn discrepancy between its predictions and certain observations on the orbit of

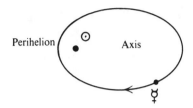

Perihelion   Axis

Figure 10. Newtonian elliptical orbit.

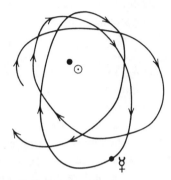

Figure 11. Precession (greatly exaggerated) of perihelion.

the planet Mercury. Now we must immediately say that Newtonian mechanics is an amazingly successful theory by any standards. Planetary orbits, eclipses, etc. can be predicted (and retrodicted) with astonishing accuracy. Using Newtonian theory, spacecraft are programmed to go back and forth to the Moon with tracking errors of only a few seconds and kilometers. As Major William Anders said during a flight to the Moon in 1968, 'I think Isaac Newton is doing most of the driving right now'. But astronomical observations are astonishingly accurate also, and it began to be realized around 1850 that the 'precession of the perihelion of Mercury' was not occurring at the predicted rate. Let us explain what this means: if there were no planets other than Mercury, its Newtonian orbit would be an ellipse, whose axes, and hence perihelion (point of closest approach) were fixed (figure 10). However, Mercury is not the only planet, and the main effect of the others is to make the axis of the orbit slowly rotate, at a rate of about 1.5° per century. This is called *precession*, and the resulting orbit traces out a rosette pattern (figure 11). The precise value of the 'centennial precession', as calculated by Newtonian mechanics, is

$$\Delta\phi_N^{100} = (5557.62 \pm 0.20)'' \, (\female\!\!\!\!\text{\female}).$$

(This value includes not only the effects of perturbations but also the effect of rotation of our Earth-based coordinate system.) The observed value is

$$\Delta\phi_{\text{obs}}^{100} = (5600.73 \pm 0.40)'' \ (\female).$$

Therefore there is a discrepancy between Newtonian theory and observation, of magnitude

▶ $$\Delta\phi^{100} \equiv \Delta\phi_{\text{obs}}^{100} - \Delta\phi_{\text{N}}^{100} = (43.11 \pm 0.45)'' \ (\female). \tag{3.2.6}$$

This discrepancy is extremely small: less than one minute of arc every century. One minute is comparable with the smallest angle resolvable by the human eye, and visualisable as 1 mm held 4 m away. Its successful detection is a triumph for the precision of Newtonian calculation and astronomical observation. The first suggested explanations postulated a new planet within Mercury's orbit, or alternatively a substantial mass of tenuous gas orbiting near the Sun. The new planet was even given the name Vulcan in anticipation of its discovery. This confidence was based on the discovery in 1846 of the planet Neptune in the outermost fringes of the solar system as it was then known. Neptune's existence had been inferred from an analysis of irregularities in the motion of Uranus. But Vulcan was not found, nor was there any convincing evidence for a massive gas cloud, despite exhaustive searching. The tentative conclusion reached by nineteenth-century astronomers was that Newton's inverse-square law might not hold precisely; this was a most unsatisfactory *ad hoc* hypothesis.

To sum up, Newton's gravitational theory cannot explain (*a*) the equality of inertial and passive gravitational mass, or (*b*) the anomalous centennial precession of Mercury's orbit.

3.3 Difficulties with Newtonian mechanics: inertial frames and absolute space

It is easy to see that Newton's second law of motion, equation (3.2.1), cannot hold in all reference frames. For the right-hand side, **F**, describes the effect on the body being considered (the 'test body') of all the other bodies in the universe, and is *invariant*, that is, independent of the coordinate frame we happen to employ to describe it. On the other hand, the left-hand side, *m***a**, involves the acceleration of the test body relative to some specified frame, and so varies from frame to frame, Thus, if we find a frame where **F** = *m***a**, we can simply choose a different frame accelerating at a rate **A** relative to the first; in this new frame the test body has an acceleration **a**' = **a** − **A**. Therefore in the new frame we no longer have **F** = *m***a**' but **F** = *m*(**a**' + **A**), which is not Newton's law.

To see this more clearly, imagine being in a space craft so far from all

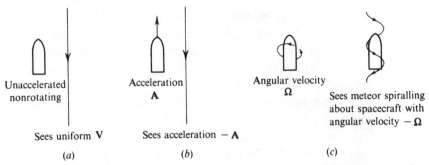

Figure 12. Motion seen from (*a*) unaccelerated, (*b*) linearly accelerated, and (*c*) rotating frames.

matter that no forces act (perhaps we are in between galaxies, or in an imaginary universe containing only ourselves and test bodies of negligible mass). Suddenly a meteor rushes by (figure 12), and we chart its course by laser ranging; it turns out to have travelled in a straight line with constant speed. No forces act on it so that by Newton's first law (or the second with $\mathbf{F} = 0$) this unaccelerated motion is exactly what we would have expected. Now suppose our spacecraft had been spinning or accelerating. Then the meteor would seem to be accelerating as it goes by, even though no forces act on it: we would have been using reference frames in which Newton's laws do not apply.

The special frames in which Newtonian mechanics can be employed are called *inertial frames* (that is, frames in which the first law – the law of inertia – holds). Near any event (place and time) there are infinitely many inertial frames, moving relatively with uniform velocities, because the acceleration of a test body with respect to a frame f is the same as the acceleration with respect to a second frame f' moving with constant velocity relative to f. Thus if Newton's law holds in f, it will hold in f' also. The idea that physical laws (e.g. Newton's laws) are the same in different inertial frames is extremely powerful; Galileo understood it perfectly, as we can see from the following quotation, from his 'Dialogues concerning two new sciences':

> *Salviatus.* Shut yourself up with some friend in the main cabin below decks on some large ship, and have with you there some flies, butterflies and other small flying animals. Have a large bowl of water with some fish it it; hang up a bottle that empties drop by drop into a wide vessel beneath it. With the ship standing still, observe carefully how the little animals fly with equal speed to all sides of the cabin. The fish swim indifferently

in all directions; the drops fall into the vessel beneath; and, in throwing something to your friend you need throw it no more strongly in one direction than another, the distances being equal; jumping with your feet together, you pass equal spaces in every direction. When you have observed all these things carefully (though there is no doubt that when the ship is standing still everything must happen in this way), have the ship proceed with any speed you like, so long as the motion is uniform and not fluctuating this way and that. You will discover not the least change in all the effects named, nor could you tell from any of them whether the ship was moving or standing still. In jumping, you will pass on the floor the same spaces as before, nor will you make larger jumps toward the stern than toward the prow even though the ship is moving quite rapidly, despite the fact that during the time that you are in the air the floor under you will be going in a direction opposite to your jump. - In throwing something to your companion, you will need no more force to get it to him whether he is in the direction of the bow or the stern, with yourself situated opposite. The droplets will fall as before into the vessel beneath without dropping toward the stern, although while the drops are in the air the ship runs many spans. The fish in their water will swim toward the front of their bowl with no more effort than toward the back, and will go with equal ease to bait placed anywhere around the edges of the bowl. Finally, the butterflies and flies will continue their flights indifferently toward every side, nor will it ever happen that they are concentrated toward the stern as if tired out from keeping up with the course of the ship, from which they will have been separated during long intervals by keeping themselves in the air. And if smoke is made by burning some incense it will be seen going up in the form of a little cloud, remaining still and moving no more toward one side than the other.

Sagredus. Although it did not occur to me to put these observations to the test when I was voyaging I am sure that they would take place in the way you describe. In confirmation of this I remember having often found myself in my cabin wondering whether the ship was moving or standing still; and sometimes at a whim I have supposed it going one way when its motion was the opposite...

However, we wish to ask deeper questions: why does Nature single out such 'preferred' frames of reference? What determines whether a given frame is inertial? Inertial frames are unaccelerated and non-rotating, but relative to what? Newton thought he had the answer: relative to *absolute space*, which 'in its own nature, without relation to anything external, remains always similar and immovable'. According to him, acceleration and rotation relative to absolute space are detectable by simple experiments. For example, if I accelerate forwards, I see the room accelerate backwards. However, there is no backward force acting on the room, so that its apparent acceleration is not a consequence of Newton's second law, but of my acceleration relative to absolute space. Another example: I see a rubber sphere rotating; is its rotation absolute, or is it a relative rotation that I perceive because, unknown to me, I am circling round it? This is easy to answer: I look and see whether the sphere bulges at the equator; if it does then the sphere is rotating relative to absolute space (like the Earth), if not, not.

And yet there are deep objections to this notion of absolute space. Firstly, how are we to identify *which* inertial frame is at rest relative to absolute space? Newton could not answer this. The second objection is subtler, but stronger: Newton's absolute space is a physical entity, it acts on matter (it is the 'seat of inertia' resisting acceleration in the absence of forces). However, matter does not act on it (it is 'without relation to anything external'), and, to quote Einstein: 'It conflicts with one's scientific understanding to conceive of a thing which acts, but cannot be acted upon.'

The third objection comes from experiment: the rigid Earth (frame 1) is a good approximation to an inertial frame for many purposes, because we can verify Newton's laws in the laboratory (figure 13). But the Earth rotates, and we get closer to an inertial frame by choosing another frame (frame 2), in which the Sun is at rest relative to the average motion of the nearby stars; this enables us to use Newtonian mechanics to describe motion in the solar system with great accuracy. But the Sun revolves about the galactic centre, and a better inertial frame (frame 3) is one in which the centre of the galaxy appears at rest relative to the average motion of other galaxies; in frame 3 the dynamics of our galaxy can be understood within Newtonian mechanics. Frame 1 rotates relative to frame 2, with angular velocity

$$\omega_{12} = 2\pi \text{ radians/day}^{-1},$$

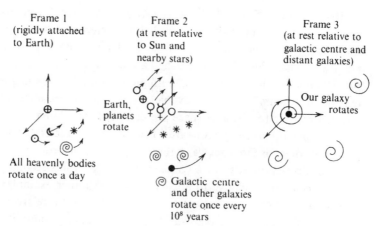

Figure 13. Successive approximations to Newtonian inertial frame.

while frame 2 rotates relative to frame 3 with angular velocity

$$\omega_{23} = 2\pi/10^8 \text{ radians year}^{-1} \sim 3 \times 10^{-11}\omega_{12}.$$

It seems that we have a rapidly converging sequence of approximations to an ideal Newtonian inertial frame. This frame is *at rest relative to the average motion of the matter in the universe*. Since the universe is expanding, an equivalent specification of this 'ultimate frame' is that from it the expansion appears isotropic. The third objection to Newton's absolute space, then, is that no explanation is given of all this experimental evidence that it seems not to be absolute at all, but somehow tied to the distribution of matter on a large scale.

3.4 Inadequacy of special relativity

In the nineteenth century the detailed unravelling of electric and magnetic phenomena culminated in Maxwell's theory of electromagnetism. This predicted that *electromagnetic waves* should exist, which travel (*in vacuo*) with a speed

$$c = 1/\sqrt{(\epsilon_0 \mu_0)},$$

where ϵ_0 and μ_0 are independent electric and magnetic quantities. When the values are inserted, c comes out as the known velocity of light, which is thus identified as an electromagnetic phenomenon. This connection of optics with electromagnetism was one of the great unifications of theoretical physics. With respect to what is the speed c? With respect to the *ether*, was the nineteenth-century answer. The ether was the medium 'in which' electromagnetic waves were supposed to travel, rather as air is the medium in which sound waves travel. It was natural to identify the ether with Newton's absolute space; this would unify electromagnetism with mechanics. Therefore experiments were carried out with the aim of discovering the Earth's velocity through the ether. Imagine the ether flowing past the Earth as a river flows past a small island; then the 'upstream' and 'downstream' speeds of light should differ. The Michelson–Morley experiment of 1887 showed that no such difference exists within the small error of the experiment. Perhaps the Earth drags nearby ether along with it, so that terrestrial experiments could not reveal any direction-dependence of the speed of light. However, this is ruled out by experiments on the aberration of starlight.

The problem of the undetectability of motion relative to the ether was solved by Einstein in 1905. He started from Galileo's relativity principle; this states that in all inertial frames the same physical laws must hold. But he rejected the 'commonsense' Galilean coordinate transformations. These concern two frames f and f', with the origin of f' moving relative to f with speed v along the positive x-axis of f; at time $t = t' = 0$, f and f' are coincident. For an event whose f coordinates are x, y, z, t, the f' coordinates are x', y', z', t', given by

$$\left. \begin{aligned} x' &= x - vt \\ y' &= y \\ z' &= z \\ t' &= t \end{aligned} \right\} \text{Galilean transformation.} \qquad (3.4.1)$$

Instead of these, Einstein boldly asserted as a postulate what had been observed experimentally: the vacuum speed of light (c) is the same in all inertial reference frames. Using this postulate, and Galileo's principle that physical laws must be the same in frames whose relative motion is uniform, he was led to the same transformation previously derived by Lorentz from Maxwell's theory:

$$\left. \begin{aligned} x' &= (x - vt)/\sqrt{(1 - v^2/c^2)} \\ y' &= y \\ z' &= z \\ t' &= (t - vx/c^2)/\sqrt{(1 - v^2/c^2)} \end{aligned} \right\} \text{Lorentz transformation.} \qquad (3.4.2)$$

From these relations follows the whole of *special relativity*, a theory abundantly verified every day in experiments with particle accelerators.

Special relativity does unify mechanics with electromagnetism, but

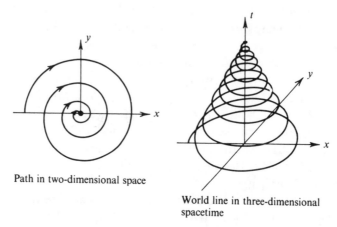

Path in two-dimensional space

World line in three-dimensional
spacetime

Figure 14. Difference between space and spacetime.

shows that if absolute space exists it cannot be identified with the ether. Nevertheless, it does not solve the problems described in sections 3.2 and 3.3; we know how to transform between inertial frames, but not what determines whether or not a given frame is inertial, nor why the frame at rest relative to the average matter in the universe is inertial. It is possible to graft gravity on to special relativity as just another force, but then the problem of the equality of m_I and m_G^P remains, as does most of the anomaly in the centennial precession of Mercury (see the end of section 5.3).

Nevertheless, we know that any successful 'general relativity' theory that does solve these problems must be consistent with special relativity; that is, it must reduce to it for appropriate limiting cases. In particular, this implies that in general relativity we must not use Newtonian absolute time together with three space coordinates, but as in special relativity we should specify events in *spacetime*. The *history* of a body moving along a path in space is summed up by a line in spacetime. This is the *world line*. of the body (figure 14). In general, space is three-dimensional, so that *spacetime is four-dimensional*. This is more than a trivial change of nomenclature, because we can specify an event by any four numbers, which need bear no resemblance to position and time coordinates (cf. the first paragraph of section 4.1), and we know in any case from special relativity that the time intervals between events are not invariant, but are different in relatively-moving frames. Thus we can see that another ingredient in a general relativity theory will be *geometry*, to enable curves in four-dimensional spacetime to be studied; this will be discussed in chapter 4.

The spacetime viewpoint was introduced into relativity by Minkowski in 1908; he said:

> The views of space and time which I wish to lay before you
> have sprung from the soil of experimental physics, and therein
> lies their strength. They are radical. Henceforth space by itself,
> and time by itself, are doomed to fade away into mere shadows,
> and only a kind of union of the two will preserve an indepen-
> dent reality.

3.5 Mach's principle, and gravitational waves

From the beginning Newton's absolute space attracted scepticism. The philosophers Leibniz and Berkeley argued that only *relative motion* could possibly be significant. Consider a universe containing only two bodies: are they rotating about their centre of mass, or not? Newton would argue that the question can be decided by joining the bodies with a spring balance; if the system had an absolute rotation the spring would extend, in order to provide the inward force necessary to keep the bodies moving in a circle, and this extension would be observable. Leibniz and Berkeley would disagree, and assert that there never would be any ex-tension of the spring. They would claim that Newton's intuition is based on his observations on the real universe, where the stars and galaxies provide a background *relative* to which non-inertial frames accelerate.

It was only a short step, taken by Mach in 1872, to assert that it is not absolute acceleration but acceleration relative to the distant matter of the universe that determines the inertial properties of matter. Thus it is not surprising that Newtonian inertial frames are found to be those which are unaccelerated relative to the average motion of the distant galaxies. There is a simple experiment making this argument very persuasive: stand outside on a starry night and look up. The stars are at rest and your hands hang down limply by your sides. Now pirouette as fast as you can. Two remarkable things happen: the stars rotate, and your arms rise up and point outwards. In Mach's view it is impossible to believe these events to be unconnected. It must be possible to develop physics in the rotating frame, in which the rotating stars exert an 'inertial force' on your arms; in other words cosmology must be related to laboratory physics.

However, Mach did not suggest *how* the stars and galaxies exert a physi-cal influence on matter near us, and to that extent 'Mach's principle' is half-baked. We can make it three-quarter-baked by the following simple argument, which is, we emphasise, not a proper general-relativistic treatment. Consider exerting a force **F** on a body; the Newtonian description is that this gives it an acceleration **a** = **F**/m relative to an inertial frame f. Now consider the body from the point of view of an accelerated, and therefore non-inertial, frame f' moving with it. In f' the body has the acceleration **a**' = 0, and by Mach's principle it must be possible to explain this by the action of forces, just as the acceleration in f was explained. What forces act in f', on the body? Clearly **F** still acts, but now there must be another force **F**', equal and opposite to **F**, so that

$$m\mathbf{a}' = 0 = \text{total force} = \mathbf{F} + \mathbf{F}'.$$

What is the force **F**'? It must come from the *acceleration of the rest of the universe*, acting on the body because this acceleration is present in f' and absent in f. The value of **F**' is

▶ $$\mathbf{F}' = -\mathbf{F} = -m\mathbf{a}, \tag{3.5.1}$$

and we have to explain how such a force can arise.

We consider first the contribution to **F**' from one mass M (say a galaxy), and then add up the contributions from all the masses in the observable universe, using cosmographical data. What force is exerted on m by M as it accelerates with **a** relative to m? First we have the ordinary static gravitational force, whose magnitude is GMm/r^2, where r is the distance of M from m. The inverse-square dependence on r means that nearby masses M are the most important; indeed, we know that it is the Earth's gravitational force on us which dominates over the effect of all other bodies. The total contribution of the static forces from the rest of the universe is essentially zero, because the pull of any distant galaxy is always counteracted by the pull of one in the opposite direction, on the average. In any case the inverse square law is not acceleration-dependent, and so cannot represent **F**'. If we demand that gravitation be at least consistent with special relativity, then it ought to resemble electromagnetism in its theoretical structure. Now the inverse-square law of gravitation is analogous to Coulomb's law between charges, but there is another force between two charges q_1 and q_2 which are mutually accelerated; its magnitude is

$$F_{\text{acc}} = q_1 q_2 a / c^2 4\pi\epsilon_0 \, r.$$

This force causes the motion of electrons in our radio aerials in response to the acceleration of electrons in the transmitter aerial. The gravitational analogue of F_{acc} is obviously the force **F**' we are looking for, and its magnitude is

▶ $$F' = GmMa/c^2 r. \tag{3.5.2}$$

The $1/r$ dependence is particularly satisfying, because it means that the distant matter contributes more significantly to $\mathbf{F}'$ than to the static gravitational force. The appearance of c^2 follows on dimensional grounds, once it is realised that c is the only constant of nature that can reasonably appear in $\mathbf{F}'$, apart from G. Sciama has called (3.5.2) the '*Law of inertial induction*' by analogy with the electromagnetic case, where the $1/r$ fields correspond to radiation.

Now we must add up the acceleration forces $\mathbf{F}'$ due to all the masses M in the universe; considering magnitudes only, for simplicity, we have

$$F' = \sum_{\text{all} M} \frac{GmMa}{c^2 r} = \frac{Gma\rho_{\text{gal}}}{c^2} \iiint_{\substack{\text{observable} \\ \text{universe}}} \mathrm{d}\mathbf{r} \frac{1}{r}$$

$$= \frac{4\pi Gma\rho_{\text{gal}}}{c^2} \int_0^{c/H} \mathrm{d}r\, r^2 \frac{1}{r}$$

(the upper limit c/H is a crude estimate of the 'horizon' at which according to Hubble's law the galaxies would be unobservable because they would be receding as fast as light). Thus

▶ $$F' = ma \times \left\{ \frac{2\pi G\rho_{\text{gal}}}{H^2} \right\}. \tag{3.5.3}$$

We already know from (3.5.1) that $F' = ma$, so that the quantity in the curly brackets should come out to be unity, if there is any truth in our crude model for Mach's principle. Let us put in the numbers for ρ_{gal} (equation (2.1.2)) and H (equation (2.3.3)); we get

$$\{ \qquad \} \approx \tfrac{1}{25}$$

instead of unity. Considering the vast range of numbers involved, this is a pretty impressive result. The main source of uncertainty lies in the value of ρ_{gal}; if it were 25 times larger, the discrepancy would disappear. However, we saw in section 2.2.4 that there might be much more matter between the galaxies than in them, so that the value of $1/25$ is a discrepancy in the right direction. (The discrepancy can be halved by increasing the masses of the distant galaxies by the relativistic factor $(1 - v^2/c^2)^{-\frac{1}{2}}$ which results from their rapid recession.)

In any case, we may expect Mach's principle to be involved in general relativity in some way: the local behaviour of matter must occur only as a result of influences from the rest of the universe, and no part of the motion must occur as a result of action by 'absolute space'.

The analogy between gravitation and electromagnetism which led to the law of inertial induction (3.5.2) suggests that *gravitational waves* should exist, which transmit with speed c effects of the acceleration of masses, just as electromagnetic waves transmit the effects of the acceleration of charges. Since cosmic objects are in constant mutual acceleration, we expect the universe to be criss-crossed with gravitational radiation. A proper mathematical treatment of this radiation is based on general relativity but is a much too sophisticated application of that theory to be dealt with by the simplified methods to be expounded later in this book. Therefore we base our discussion on (3.5.2).

How are gravitational waves produced? How intense are they? How could we detect them? We attack these questions with the aid of an example, in which the source of gravitational waves is the rotational acceleration of the components of a binary star. To get the strongest possible radiation we place the stars at a distance 1 pc from us (roughly the distance of the nearest stars), and let them be neutron stars with a radius $R = 6$ km and mass equal to that of the Sun, so close that their surfaces are in contact. Then their mutual acceleration is

$$a = GM_\odot /(2R)^2 \approx 9 \times 10^{11} \,\text{m s}^{-2} \approx 10^{11} \,g.$$

The resulting force F' on a test mass m in the vicinity of the Earth is given by (3.5.2), and the acceleration A of this mass is

$$A = G(2M_\odot)a/c^2r \approx 0.09 \,\text{m s}^{-2} \approx 10^{-2} g.$$

This is rather large (and 2×10^{11} larger than the acceleration due to the inverse-square static force). Nevertheless it would be hard to detect A directly (by for example the resulting change in the rate of a pendulum clock). The reason is that A is not steady but oscillatory, the wave period being half the rotation period T of the binary (if the stars are identical). Thus the period is

$$T/2 = 2\pi\sqrt{(R^3/GM_\odot)} \approx 10^{-4}\,\text{s} \approx 10^{-4}\,\text{s}$$

so that we are dealing with 10 kHz gravitational radiation. To get a resonant response we would need a pendulum with the same frequency, but the length of this would be only about 20 Å.

The gravitational wave also varies in space (with wavelength $cT/2$); test masses half a wavelength apart move in antiphase, and the possibility arises of measuring the relative velocity V and displacement X. The size of a 'half-wave aerial' is

$$cT/4 \approx 10 \,\text{km}.$$

If the wave is harmonic, the maximum oscillation speed V is

$$V = A/\text{angular frequency} = A(T/2)/2\pi \approx 3 \times 10^{-6} \,\text{m s}^{-1}.$$

This gives a Doppler shift of $v/c = 10^{-14}$, which is on the borderline of detectability using the Mössbauer effect. The amplitude X of the oscillation is

$$X = A(T/2)^2/(2\pi)^2 \approx \tfrac{1}{2} \text{ Å},$$

which is again barely detectable.

A range of different gravitational-wave-generating mechanisms has been examined theoretically (e.g. matter accelerating into a black hole (see section 5.6) or the explosion of a supernova), and a variety of ingenious detecting devices have been designed and built. As yet, however, no gravitation waves have been observed with certainty (several claims to such observation have been made, but not generally accepted). If over the next decades improvements in high-precision technology result in detectors a few orders of magnitude more sensitive than those available today, then we can look forward to the beginning of *gravitational wave astronomy*. This could profoundly affect our understanding of the universe. It would correspond to a radical extension of our mode of perception of cosmic events: radio waves and X-rays (being electromagnetic radiation) extend the way we *see* the universe; cosmic rays (being particles from space) enable us (metaphorically) to *smell* the universe; gravitational radiation (actually shaking matter on Earth) will enable us to *feel* the pulse of the universe.

3.6   Einstein's principle of equivalence

Suppose we have a region of spacetime where a constant gravitational field $\mathbf{g}$ acts (i.e. the gravitational force on a particle of mass m is $m\mathbf{g}$). One such region is the neighbourhood of any point on the surface of the earth. Then, if gravity were the only force acting, all bodies in the region would fall with the same acceleration $\mathbf{a} = \mathbf{g}$ (since $m\mathbf{a} = m\mathbf{g}$). Therefore, by transforming to a frame f' with acceleration $\mathbf{g}$ we can eliminate the effects of gravitation; any body will appear unaccelerated unless a non-gravitational force $\mathbf{F}_{ng}$ acts on it. It is easy to show this formally: in the original frame f, Newton's law gives

$$m\mathbf{a} = m\mathbf{g} + \mathbf{F}_{ng}.$$

In f', the acceleration is $\mathbf{a}' = \mathbf{a} - \mathbf{g}$, while the same forces act. Therefore

$$m(\mathbf{a}' + \mathbf{g}) = m\mathbf{g} + \mathbf{F}_{ng},$$

that is

▶ $$m\mathbf{a}' = \mathbf{F}_{ng}. \tag{3.6.1}$$

In this equation of motion, gravitational forces do not enter. Now the frame f' is *in free fall* in the gravitational field. Thus the equality of

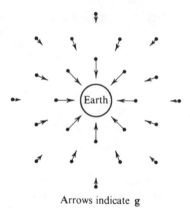

Arrows indicate **g**

Figure 15. Inhomogeneous gravitational field.

gravitational and inertial mass implies the following:

> In a small laboratory falling freely in a gravitational field,
> mechanical phenomena are the same as those observed in a
> Newtonian inertial frame in the absence of a gravitational field.

In 1907 Einstein generalised this conclusion, replacing 'mechanical phenomena' by 'the laws of physics'; the resulting statement is the *principle of equivalence*.

Why must the laboratory be 'small'? Because real gravitational fields are only locally constant: **g** points towards the earth's centre, and so its direction varies from point to point round the earth; also its magnitude varies with height above the earth (figure 15). Now, a frame of reference involves a rigid arrangement of coordinate rods, and so can only fall with one single acceleration. Therefore only the centre of mass of a 'large' laboratory is in free fall. This could in principle be detected as follows: consider a laboratory released from rest outside the Earth, and containing two particles at the same distance from the Earth. The laboratory, and the particles, will fall towards the Earth's centre. The particles will move along radii, (figure 16), so that an observer in the laboratory will see them approach one another, as, unknown to him, the laboratory nears the earth. However, it is easy to show that the relative speed with which the particles approach one another is given by

$$u = (d/R^2) \sqrt{(2GMh)},$$

that is

▶ $$u = (d/R) \sqrt{(2gh)} \tag{3.6.2}$$

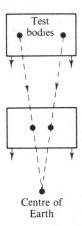

Figure 16. Einsteinian laboratory must be small.

where d is their separation, R their distance from the Earth's centre, and h the distance fallen. This is very small; starting 10 m apart and falling 10 km near the Earth's surface, the formula gives $u \sim 10^{-3}$ m s^{-1}, which is detectable but small. In any case, u is proportional to d, so the smaller d is (i.e. the smaller the laboratory), the less easy it is to observe the approach of the particles, and the harder it is to infer the presence of the external gravitating body.

These freely-falling frames covering the neighbourhood of an event are very important in relativity; they are called *local inertial frames*. They are not inertial in the strict Newtonian sense, because the equation of motion (3.6.1) only corresponds to Newton's second law if the gravitational force $m\mathbf{g}$ is ignored. But the existence of this force cannot be established anyway (at least, not by experiments within the frame), so that Einstein's notion of an inertial frame is closer to what Newton intended. Near any event there is an infinity of local inertial frames, each moving with constant velocity relative to the others. Special relativity applies rigorously within these frames, and the Lorentz transformation tells us how to transform the coordinates of an event from one of these frames to another. Local inertial frames are at once more restricted and more general than Newton's inertial frames: more restricted, because the inhomogeneity of real gravitational fields makes them only locally applicable, instead of infinite in extent; and more general, because *any* laboratory in free fall (for instance *Skylab*) is a local inertial frame – it does not have to be unaccelerated relative to the galaxies, or to 'absolute space'.

Thus we can 'transform away' the *local* effects of a gravitational field by

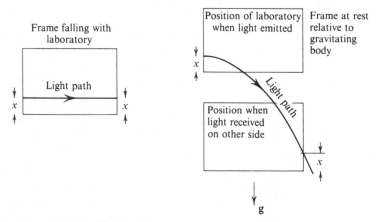

Figure 17. Principle of equivalence predicts 'fall' of light.

employing a laboratory in free fall, but no single laboratory exists which can cover all the space round a gravitating object and move in such a way as to eliminate its effects. This is obvious, because 'tidal effects' of gravitation, arising from the difference between the values of **g** at different points, are real. There is only one exception to this: if there were no gravitating bodies, there would be no 'free fall', and the local inertial frames would in fact extend to infinity; special relativity would apply everywhere. In real cases we have to cover spacetime with a patchwork of local inertial frames, and we shall see in chapter 4 how general relativity enables us to carry out the patching.

The principle of equivalence leads very simply to two testable conclusions about the propagation of light. The first is that *light bends in a gravitational field*. Consider a freely-falling laboratory, for example a spacecraft, or a lift whose supporting cables have been cut. Let a light ray be emitted from one side of the laboratory (figure 17) in a direction perpendicular to the local gravitational field direction. The principle tells us that to an observer inside the laboratory the light travels according to the laws of the special theory of relativity, that is, in a straight line with speed c. But the laboratory is accelerating downwards, so that as seen from outside the light accelerates downwards too; that is, it pursues a curved path. Near the Earth's surface, where **g** is locally constant and parallel, this 'fall' of light is far too small to be measurable; a horizontally-projected beam, after travelling 1 km, has fallen only about 1 Å. It is, however, possible to detect the deflection of light falling past the Sun, but this involves the patching-together of many inertial frames, and we put off the calculation until section 5.4. We may say now, however, that experiment supports the deflection predicted by general relativity, which is twice as great as that predicted on the basis of Newtonian mechanics by treating light as a stream of particles starting out with speed c.

The second testable conclusion is the *gravitational shift of spectral lines*. This has already been introduced in section 2.3, where we considered the energy loss of a photon 'climbing' out of a gravitational field. The equivalence principle entered that derivation in a disguised form when we used the 'inertial mass' $h\nu_0/c^2$ as the passive gravitational mass in the Newtonian work formula. Now we derive the spectral shift formula in an alternative way, which employs the principle directly. As a laboratory of height h is released from rest into free fall, let light of frequency ν_e be emitted from the floor and travel upwards. By the principle of equivalence, it hits the ceiling after a time $t = h/c$, and to an observer in the ceiling the light still has frequency ν_e. But an observer outside the falling laboratory is now moving upwards (away from the light) relative to the laboratory, with speed $u = gt = gh/c$. Therefore this outside observer sees the light Doppler-shifted towards the red, by an amount given by the usual classical formula

▶ $$z = u/c = gh/c^2. \tag{3.6.3}$$

The quantity gh is the change in Newtonian gravitational potential experienced by the light; thus this formula is essentially the same as (2.3.4), which contains the analogous quantity GM_g/r_g, the change in potential experienced by light escaping from r_g to infinity. For light falling rather than climbing, (3.6.3) gives a negative z, that is, a blue shift. The argument we have employed is interesting, because it shows that by choosing a freely-falling local inertial frame in which to analyse the motion of light, a gravitational shift appears as a Doppler shift. This is a practical example of how a gravitational field can be locally 'transformed away'.

The gravitational shift can be measured in terrestrial laboratories. In 1960 Pound and Rebka allowed a 14.4 keV γ-ray, emitted by the radioactive decay of ^{57}Fe, to fall 22.6 m down an evacuated tower, and they measured the change in its frequency. The predicted blue shift is $z = -2.47 \times 10^{-15}$, and they measured $z = (-2.57 \pm 0.26) \times 10^{-15}$, thus directly verifying the equivalence principle. Such incredible precision was possible because of the Mössbauer effect; this is the emission of radiation from an atomic nucleus in a crystal, which gives a spectral line with a very precisely-defined frequency.

There is an alternative explanation of these spectral shifts. A radiating atom may be regarded as a clock, each 'tick' being the emission of a wave crest. Since light from a clock in a gravitational field will be received reddened by a 'clock at infinity', the latter clock will see the former clock ticking more slowly than itself if the clocks are of identical construction. The gravitational time dilation factor for a clock distant r from a mass M, is

▶ $$\Delta t(r)/\Delta t(\infty) = 1 - GM/rc^2, \tag{3.6.4}$$

where $\Delta t(r)$ and $\Delta t(\infty)$ are time intervals between events, measured in terms of ticks of the clocks at r and infinity.

4 Curved spacetime and the physical mathematics of general relativity

4.1 Particle paths and the separation between events

Now we must start to get to grips with general relativity itself. We wish to predict the motion of a body under the action of gravitation, avoiding the difficulties already described which result from the Newtonian formalism. Therefore we employ a *spacetime* description, and seek the *world line* of the body, that is, the locus of successive *events* in its history. Each event is a point in spacetime, specified by four *coordinates* x^i, where $i = 0, 1, 2, 3$. It is conventional to take x^0 as a time coordinate t, and x^1, x^2 and x^3 as three coordinates giving the spatial position $\mathbf{r}$ ($= x, y, z$; r, θ, ϕ etc.). We have in mind here a lattice formed by rigid 'standard rods'. whose intersections give spatial positions; at each intersection is a 'standard clock' which can be synchronised with all the others by means of a technique involving light signals – it is not necessary for us to go into details here. For any event, the nearest intersection gives $\mathbf{r}$, and the clock reading there gives t. However, much more general specifications of events can be employed. For example, an explosion in the atmosphere can be specified by taking the readings of clocks carried by four randomly-moving aeroplanes, at the moment the explosion is seen or heard; the clocks need not be synchronised and may even run at different speeds, but they must keep going. The four readings thus obtained specify the explosion event uniquely, and form a perfectly acceptable set of coordinates x^i. This freedom in choosing coordinate systems is particularly useful in view of the fact that a rigid body, such as a 'standard rod', is not easy to define in relativity (the trouble arises because the speed of sound in such a body would be infinite, and therefore greater than that of light). However, we shall often speak about 'coordinate lattices', relying on the result that an equivalent 'coordinatisation' of spacetime can be carried out using light rays alone (see the more advanced texts in the bibliography).

Einstein argued that physical events happen independently of our observations (except for quantum mechanical effects which are negligible

[47]

cosmically); therefore it must be possible to express physical laws in equations which take the same form whatever coordinates x^i we choose to describe the events the laws set out to explain. Such equations are called *covariant*, and Einstein's principle is the *principle of general covariance*. Newton's laws, and the equations of special relativity, are not generally covariant, because they hold only in inertial frames: the coordinate lattice must not accelerate or rotate. These theories do possess a restricted covariance: they can be written with positions, velocities and accelerations in *vector* form, so that within the permitted inertial frames any *spatial* coordinates may be employed, such as Cartesian, polar, cylindrical etc. However, we are not now discussing spatial covariance, but covariance in spacetime. Einstein succeeded in writing a generally covariant set of gravitational equations; mathematically, they are very complicated. Fortunately, this full generality is not required in most cosmological applications of relativity, since the coordinate systems x^i commonly employed involve a clear separation between spatial and time coordinates.

How can we describe the world line of a body? In Newtonian mechanics the sensible procedure is to use the (absolute) time coordinate as a parameter, and express the spatial position **r** as a function of t; thus we would write **r**(t). In relativity, however, this procedure is unsymmetrical, because t is no longer an absolute parameter, but just another coordinate, namely x^0. We would like an absolute parameter τ which increases smoothly along a particle's world line from its past to its future; then a 'covariant' specification of the history of the particle would be the four functions $x^i(\tau)$. The natural parameter τ is the *proper time*; this is the time read by a standard clock travelling with the body (for example, the 'ticks' might be successive crests of light waves emitted during a certain transition between energy levels of the atoms of the body). Obviously τ is an *invariant* parameter (it has been introduced without mentioning any coordinate system).

To clarify this use of the parameter τ to specify world lines in spacetime, we examine the familiar simpler case of specifying curves in the ordinary Euclidean three-dimensional space of positions. There a point is described by three coordinates x^μ where $\mu = 1, 2, 3$ (figure 18). Curves are parametrised by using the *arc length* s, that is, they are specified by the three functions $x^\mu(s)$. On any curve, the arc length Δs between neighbouring points x^μ and $x^\mu + \Delta x^\mu$ is given by Pythagoras' theorem. In Cartesian coordinates x, y, z, this gives

▶ $$\Delta s^2 = \Delta x^2 + \Delta y^2 + \Delta z^2. \tag{4.1.1}$$

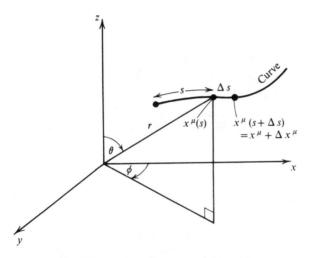

Figure 18. Arc length s parameterises a space curve.

An alternative representation of the same Δs in polar coordinates r, θ, ϕ is

$$\blacktriangleright \qquad \Delta s^2 = \Delta r^2 + r^2(\Delta\theta^2 + \sin^2\theta\,\Delta\phi^2). \qquad (4.1.2)$$

The analogous formula in spacetime gives $\Delta\tau^2$ as a function of the coordinate differences Δx^i between any two neighbouring events. By choosing any freely falling (locally inertial) frame near whose origin the events occur, we can employ Cartesian coordinates $x^i = t, x, y, z = t, \mathbf{r}$. Then $\Delta\tau^2$ is given by the formula from special relativity:

$$\blacktriangleright \qquad \Delta\tau^2 = \Delta t^2 - (\Delta x^2 + \Delta y^2 + \Delta z^2)/c^2 = \Delta t^2 - |\Delta\mathbf{r}|^2/c^2. \qquad (4.1.3)$$

In assuming that special relativity holds in the freely-falling frame, we have used the principle of equivalence. Actually equation (4.1.3) is very powerful; it can be made the basis of special relativity, and the Lorentz transformations, etc. derived from it. To illustrate this, we derive the *time dilation formula*. Consider two events on the world line of a body moving with speed v relative to a coordinate frame (figure 19): the coordinate time difference Δt and distance travelled $|\Delta\mathbf{r}|$ are related by $|\Delta\mathbf{r}| = v\Delta t$, so that the proper time (i.e the time the body's own clocks show) is $\Delta\tau^2 = \Delta t^2 - v^2\Delta t^2/c^2$, and

$$\blacktriangleright \qquad \Delta t = \Delta\tau/\sqrt{(1 - v^2/c^2)}. \qquad (4.1.4)$$

Thus coordinate time intervals Δt, measured in a frame moving relative to

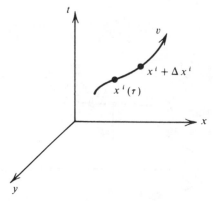

Figure 19. Two events on a body's world line.

the body, are longer than proper time intervals $\Delta\tau$, measured with clocks always at rest relative to the body. In short: *moving clocks run slow*.

There is a very important difference between the formula for $\Delta\tau^2$ in special relativity and the formula for Δs^2 from Pythagoras' theorem: between any two neighbouring points in Euclidean three-dimensional space, the distance is positive, and so, of course, is Δs^2. By contrast, it is easy in spacetime to find two neighbouring events whose 'proper time' $\Delta\tau$ is *imaginary*, that is, $\Delta\tau^2$ is negative. From (4.1.3), we see that any pair of events for which

$$|\Delta\mathbf{r}| > c\Delta t$$

has this character. A material particle joining such events would have to travel with a coordinate speed exceeding c, which is impossible; therefore these events cannot lie on the world line of any particle, and $\Delta\tau$ cannot be interpreted as a proper time. However, the positive quantity $c\sqrt{(-\Delta\tau^2)}$ does have a physical significance: it is the *proper distance* between the events. This is defined as the distance measured with a standard rod or tape, in a reference frame where the events occur simultaneously. To show this, we simply choose the local inertial frame in which $\Delta t = 0$ (simultaneity); then, from (4.1.3),

$$|\Delta\mathbf{r}|^2 = -c^2\Delta\tau^2,$$

and the stated result follows, since in this frame $|\Delta\mathbf{r}|$ is proper distance.

With this identification of $\Delta\tau$ in terms of proper distance, it is possible to derive the *Lorentz contraction formula*. Let the original reference frame be called f, and consider a second frame f' moving over f with speed v (figure 20). Let the two events being considered be the coincidence of the two

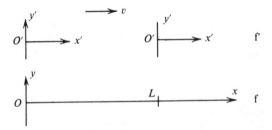

Figure 20. Relatively-moving frames.

origins O and O', and the passage of O' over the point on the x axis of f whose distance from O is L.

Then, in f, we have, for these events,

$$\Delta\tau^2 = \Delta t^2 - |\Delta\mathbf{r}|^2/c^2 = L^2/v^2 - L^2/c^2.$$

In f' both events occur at the origin O', and also the distance between the two points on the x axis will be some unknown length L', so that

$$\Delta\tau^2 = \Delta t'^2 - |\Delta\mathbf{r}'|^2/c^2 = L'^2/v^2 - 0.$$

We can equate these two expressions, and we find that

▶ $L' = L\sqrt{(1 - v^2/c^2)},$ (4.1.5)

so that rods appear contracted ($L' < L$) if their lengths are inferred from measurements in a frame in which they are not at rest.

It is also possible to have pairs of events for which $\Delta\tau$ is zero; then the coordinate differences are related by

$$|\Delta\mathbf{r}|/\Delta t = \pm c,$$

so that the events could lie on the world line of a light ray.

For any pair of events, $\Delta\tau$ is called the *separation* (sometimes the term *interval* is used instead). We can distinguish three kinds of separation:

(i) If $\Delta\tau$ is *real*, the events have a *timelike* separation, and can lie on the world line of a material particle; $\Delta\tau$ is the proper time between the events.

(ii) If $\Delta\tau$ is *imaginary*, the events have a *spacelike* separation, and cannot lie on the world line of any particle; $c\sqrt{(-\Delta\tau^2)}$ is the proper distance between the events.

(iii) If $\Delta\tau$ is *zero*, the events have a *lightlike* separation, and can lie on the world line of a light ray. The events in this case are said to have a *null* separation.

Physically, therefore, separation in spacetime is more complicated than distance in space. Mathematically, however, the two concepts are very

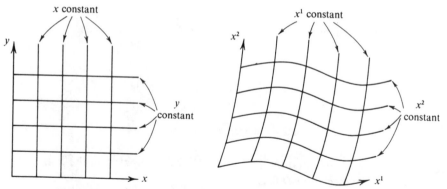

Figure 21. Cartesian and general coordinates.

close, and we can think of $\Delta\tau^2$ as being given by (4.1.3) in terms of a 'pseudo-Pythagorean' theorem where the signs of three of the four terms are negative.

Although we have promised not to use general coordinate systems, it is important to find general expressions for the distance Δs in position space, and the separation $\Delta\tau$ in spacetime. The argument is best introduced in the three-dimensional case. Suppose we locate a point by general co-ordinates corresponding to an arbitrary coordinate grid in this space (figure 21). Then if the original Cartesian x, y and z are arbitrary functions of the new coordinates x^1, x^2 and x^3 we can write this as

▶ $$x = x(x^1, x^2, x^3); \quad y = y(x^1, x^2, x^3); \quad z = z(x^1, x^2, x^3). \quad (4.1.6)$$

The coordinate differences appearing in (4.1.1) are therefore given by

$$\Delta x = \frac{\partial x}{\partial x^1}\Delta x^1 + \frac{\partial x}{\partial x^2}\Delta x^2 + \frac{\partial x}{\partial x^3}\Delta x^3,$$

$$\Delta y = \frac{\partial y}{\partial x^1}\Delta x^1 + \frac{\partial y}{\partial x^2}\Delta x^2 + \frac{\partial y}{\partial x^3}\Delta x^3,$$

$$\Delta z = \frac{\partial z}{\partial x^1}\Delta x^1 + \frac{\partial z}{\partial x^2}\Delta x^2 + \frac{\partial z}{\partial x^3}\Delta x^3.$$

Substituting into (4.1.1), we get

$$\Delta s^2 = \left[\left(\frac{\partial x}{\partial x^1}\right)^2 + \left(\frac{\partial y}{\partial x^1}\right)^2 + \left(\frac{\partial z}{\partial x^1}\right)^2\right](\Delta x^1)^2$$

$$+ 2\left[\frac{\partial x}{\partial x^1}\frac{\partial x}{\partial x^2} + \frac{\partial y}{\partial x^1}\frac{\partial y}{\partial x^2} + \frac{\partial z}{\partial x^1}\frac{\partial z}{\partial x^2}\right]\Delta x^1\,\Delta x^2 + \dots$$

$$= \sum_{\mu=1}^{3}\sum_{\nu=1}^{3} g_{\mu\nu}(x^1, x^2, x^3)\,\Delta x^\mu\,\Delta x^\nu,$$

where

$$g_{\mu\nu} = \left(\frac{\partial x}{\partial x^\mu} \frac{\partial x}{\partial x^\nu} + \frac{\partial y}{\partial x^\mu} \frac{\partial y}{\partial x^\nu} + \frac{\partial z}{\partial x^\mu} \frac{\partial z}{\partial x^\nu} \right).$$

Finally, we introduce the useful shorthand of *Einstein's summation convention*: two identical indices in any expression are to be summed over. Both μ and ν occur twice in the formula for Δs^2, so we can write it as

▶ $\qquad \Delta s^2 = g_{\mu\nu}(\mathbf{r}) \, \Delta x^\mu \, \Delta x^\nu.$ (4.1.7)

This expression is very important: it tells us how to obtain a physically significant quantity – the distance between two points – from a knowledge of the coordinate differences between the points. The distance Δs is invariant, but the coordinate differences are not; they depend on the arbitrarily chosen coordinate grid system. The formula involves the functions $g_{\mu\nu}(\mathbf{r})$; there are nine of these:

$$g_{\mu\nu} = \begin{pmatrix} g_{11} & g_{12} & g_{13} \\ g_{21} & g_{22} & g_{23} \\ g_{31} & g_{32} & g_{33} \end{pmatrix},$$

but only six are independent, since $g_{\mu\nu} = g_{\nu\mu}$. The set of functions $g_{\mu\nu}$ is called the *metric tensor*.

Tensors in an n-dimensional space are sets of functions which transform between coordinate systems in a certain way. A tensor of zero rank is a single function of position, whose value is the same in all coordinate systems; in more familiar terms, it is a *scalar*. A tensor of rank one is a set of n functions; in more familiar terms, it is a *vector*, and the n functions are its components. In the present example, the three coordinate differences $\Delta x^\mu = (\Delta x^1, \Delta x^2, \Delta x^3)$ form a vector, often written $\Delta \mathbf{r}$. A tensor of the second rank is a set of n^2 functions, and the metric tensor $g_{\mu\nu}$ is an example. The algebra and calculus of tensors form the basic mathematical structure of general relativity, and give that theory much of its power, as well as its difficulty. We shall not base our exposition on tensors because, as we have said, we shall not need the most general formulation of Einstein's theory. It is also necessary to remark that there is a technical reason why μ and ν appear as suffixes in $g_{\mu\nu}$ and superscripts in x^μ and x^ν. For further details, the reader is referred to the excellent text *Tensor calculus* by Synge and Schild (see bibliography).

The metric tensor puts the 'metry' into geometry; it is quite possible to have a general space with coordinates – a 'manifold' – which has no metric tensor. Most graphs are of this kind. For instance, in a graph showing the

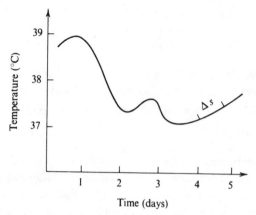

Figure 22. Non-metric space.

temperature of a hospital patient as a function of time (figure 22), the arc length Δs has no meaning; we are not dealing with a metric space.

The simplest form of the metric tensor occurs if rectangular Cartesian coordinates are used. $g_{\mu\nu}$ is then *diagonal* (i.e. $g_{\mu\nu} = 0$ if $\mu \neq \nu$) and the diagonal elements g_{11}, g_{22} and g_{33} are all unity; we write this special case as $g^0_{\mu\nu}$, and direct comparison of (4.1.1) and (4.1.7) gives

$$\blacktriangleright \quad g^0_{\mu\nu} = \begin{pmatrix} 1 & 0 & 0 \\ 0 & 1 & 0 \\ 0 & 0 & 1 \end{pmatrix}. \tag{4.1.8}$$

The argument of the last few pages may be applied in a precisely analogous manner to *spacetime*: in arbitrary, possibly accelerating and rotating reference frames the separation $\Delta\tau$ between two events is given in terms of their coordinate differences by

$$\blacktriangleright \quad \Delta\tau^2 = g_{ij}\Delta x^i \Delta x^j \equiv \sum_{i=1}^{4} \sum_{j=1}^{4} g_{ij}\Delta x^i \Delta x^j. \tag{4.1.9}$$

The metric tensor g_{ij} has sixteen components, of which only ten $((16-4)/2+4)$ are independent. The simplest form for g_{ij} holds in a freely-falling frame, if we use time for x^0 and Cartesian space coordinates for x^1, x^2, x^3; then the metric tensor is g^0_{ij}, given according to (4.1.3) and (4.1.9) by

$$\blacktriangleright \quad g^0_{ij} = \begin{pmatrix} 1 & 0 & 0 & 0 \\ 0 & -1/c^2 & 0 & 0 \\ 0 & 0 & -1/c^2 & 0 \\ 0 & 0 & 0 & -1/c^2 \end{pmatrix}. \tag{4.1.10}$$

In general, however, the components of g_{ij} are functions of the event coordinates x^i.

There are two important differences, apart from the number of dimensions, between Euclidean three-dimensional space and four-dimensional spacetime. The first difference is that in spacetime it is possible to find non-coincident events with zero separation, so the distinction can be made between spacelike and timelike separations. Mathematically, we see from (4.1.10) and the discussion based on (4.1.3) that this distinction arises from the occurrence in g_{ij}^0 of negative signs. (More precisely, it is always possible to transform an arbitrary g_{ij} to diagonal form, and find the difference between the numbers of positive and negative elements; this is called the *signature* of the metric, and its value for spacetime, as can be seen from (4.1.10), is -2.) Metric tensors for which $\Delta\tau^2$ may be zero are called *indefinite*. By contrast, the metric $g_{\mu\nu}$ of position space is *definite* (cf. (4.1.1), (4.1.2), (4.1.8)). It can be shown that these properties are invariant, that is, an indefinite metric will remain so under arbitrary real coordinate transformations (this is almost obvious).

The second difference is that while in Euclidean three-dimensional space $g_{\mu\nu}$ can always be reduced to $g_{\mu\nu}^0$ (equation (4.1.8)) by an appropriate coordinate transformation, it is generally possible in spacetime to reduce g_{ij} to g_{ij}^0 *only locally*, according to the principle of equivalence. This is merely a mathematical restatement of our physical principle that in a gravitational field it is not possible to cover the whole of spacetime with a single inertial frame. In some way, therefore, the metric tensor g_{ij} in the presence of gravity must describe a spacetime intrinsically more complicated than that of special relativity, whose metric is given by (4.1.10). We shall soon see that the new property introduced by a gravitational field is *curvature*.

4.2   Geodesics

What path $x^i(\tau)$ does a body follow, if no non-gravitational forces act? Let us first answer the question in special relativity: the body moves with uniform speed along a straight line, so that its world line as measured in an inertial frame is given by the four equations

$$\blacktriangleright \qquad \frac{d^2t}{d\tau^2} = 0; \qquad \frac{d^2\mathbf{r}}{d\tau^2} = 0. \tag{4.2.1}$$

We have simply restated Newton's first law. In a non-inertial frame, however, the $\mathbf{r}$ and t coordinates get mixed up and these equations take a different form; in any case the frame can only be *locally* inertial when an

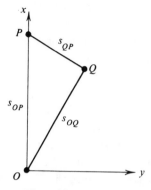

Figure 23

inhomogeneous gravitational field exists. For these two reasons we know that the equations (4.2.1) do not have the generally covariant form required for a physical law. However, we do not need to write down the equation of a straight line in order to specify it; there is another way. In Euclidean three-dimensional space, a straight line corresponds to the *shortest distance between two positions*. Let us show this: consider the origin O and the point P whose coordinates are x_P, 0, 0 (figure 23). Then the distance s_{OP} is given by Pythagoras' theorem:

$$s_{OP}^2 = x_P^2 + 0 + 0,$$
$$s_{OP} = |x_P|.$$

Now consider an alternative path OQP, where Q, not on OP, has co-ordinates x_Q, y_Q, 0. Then

$$s_{OQP} = s_{OQ} + s_{QP} = \sqrt{(x_Q^2 + y_Q^2)} + \sqrt{[(x_P - x_Q)^2 + y_Q^2]}.$$

But we can write

$$s_{OP} = |x_Q + (x_P - x_Q)|,$$

so that it is obvious that

$$s_{OQP} > s_{OP}. \quad \text{Q.E.D.}$$

This argument is general, because any point P can be made to lie on the x axis, and any curved path can be made up of segments like OQ and QP, and the limit taken.

All this seems obvious enough, but in the spacetime of special relativity the timelike separation between two events is *longest* along the straight world line joining them. Using an analogous argument, we choose our two

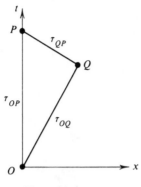

Figure 24

events as the origin O of an inertial frame and the event P whose coordinates are $t_P, 0, 0, 0$ (figure 24). This is always possible, because we can always choose a frame whose spatial origin moves with a body on whose world line O and P occur. The separation along the straight world line is, by (4.1.3),

$$\tau_{OP}^2 = t_P^2 - 0 - 0 - 0,$$

i.e. $\tau_{OP} = t_P.$

Now consider an alternative world line OQP, where Q has coordinates $t_Q, x_Q, 0, 0$. Then

$$\tau_{OQP} = \tau_{OQ} + \tau_{QP};$$

this follows from the additivity of clock readings and the necessity for OQ and QP to be timelike in order that OQP can be a possible world line. Now we use (4.1.3) again, to obtain

$$\tau_{OQ}^2 = t_Q^2 - x_Q^2/c^2;$$

and $\tau_{QP}^2 = (t_P - t_Q)^2 - x_Q^2/c^2.$

But we can write

$$\tau_{OP} = t_Q + (t_P - t_Q)$$

so that it is obvious that

$$\tau_{OQP} = \sqrt{(t_Q^2 - x_Q^2/c^2)} + \sqrt{[(t_P - t_Q)^2 - x_Q^2/c^2]} < \tau_{OP}. \quad \text{Q.E.D.}$$

This argument brings out an important incidental point: the separation between two events depends on the world line between them. In other words, if two clocks set out synchronised from one event, and travel along different world lines, they will not generally read the same times when they

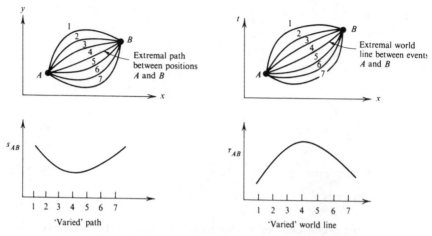

Figure 25. Geodesic is shortest distance in space, longest separation in space-time.

meet again at another event. This has occasionally been thought to be logically impossible, and given the name 'twin paradox' (the clocks are twins and the 'ticks' are, say, heartbeats). The effect has now been observed (see section 5.2), and is really no more paradoxical than the fact that two pieces of string joining the same two points can have different lengths. For future reference, we write the expression for the separation τ_{AB} between two events A and B, along a world line $x^i(\tau)$ (not necessarily straight) joining them; from (4.1.9), we have

$$
\blacktriangleright \quad \tau_{AB} = \int_A^B \sqrt{(g_{ij}\,\mathrm{d}x^i\,\mathrm{d}x^j)}
$$

$$
= \int_A^B \sqrt{\left[g_{ij}(x^k(\tau)) \frac{\mathrm{d}x^i(\tau)}{\mathrm{d}\tau} \frac{\mathrm{d}x^j(\tau)}{\mathrm{d}\tau} \right]}\,\mathrm{d}\tau. \qquad (4.2.2)
$$

It appears, then, that a straight line may correspond either to a minimum or to a maximum of some quantity, depending on the metric, that is, on the formula for Δs^2 or $\Delta\tau^2$. Both cases are covered by saying that for straight lines the quantities are *extremal*. This means that if we choose other paths between the same events or points, infinitesimally close to the extremal paths, the distance s or separation τ will have the same value as on the extremal path (figure 25). In general spaces defined by a metric tensor $g_{\mu\nu}$ or g_{ij}, these extremal paths are called *geodesics*. Where g_{ij} is indefinite, as in spacetime, geodesics along which the separation is zero may exist; they are called *null geodesics*.

The concept of geodesics is what we have been working towards. To say: 'a body pursues a timelike geodesic in spacetime' involves no particular coordinate system. It is a 'generally covariant' statement and therefore qualifies as a possible physical law. However, we have proved only that it holds within special relativity, that is, either everywhere in the absence of gravity or, if an inhomogeneous gravitational field is present, only locally. Einstein, with another characteristic stroke of boldness, postulated that the law holds generally. Thus we can sum up this section, and the previous one, in the following way:

> Spacetime is a four-dimensional manifold with indefinite metric g_{ij}, in terms of which the separation between events is given by
>
> ▶ $\Delta\tau^2 = g_{ij}\Delta x^i \Delta x^j$.
>
> The world lines of material bodies not subjected to non-gravitational forces are timelike geodesics. The world lines of light rays are null geodesics.

These three statements, together with the definition of separation in terms of rods and clocks, constitute a large part of general relativity. But an important question remains: How is the metric tensor g_{ij} determined? We shall answer this in section 4.4. Meanwhile we must discuss the geodesic law of motion a little more. It is hardly possible to imagine a more beautiful or simple description. We are asserting that spacecraft, planets, stars, and galaxies, as they trace out their complicated paths, are doing nothing more than making the separation take an extremal value between any two events on their world lines. Moreover, we have solved some of our previous problems: inertial and passive gravitational mass do not appear in general relativity, and the local constancy of g_{ij} means that it is always possible to set up a local inertial frame, where the metric tensor is given by (4.1.10); thus the principle of equivalence is built into the theory from the start. The geodesic postulate is very different in appearance from Newton's second law (3.2.1), yet its predictions agree very well with experiment; this will be shown in chapter 5. We have presented the geodesic law as an independent postulate of general relativity, but it is worth pointing out that it can in fact be derived from the other postulates of the theory (either from the principle of equivalence plus general covariance, or from the 'field equations' determining g_{ij} – these will be discussed briefly in section 4.4).

It is important to realise that the geodesic law applies to world lines between events in spacetime, and *not* to paths between positions in ordinary three-dimensional space. The latter alternative would be nonsense, because any two positions A and B can be joined not just by one path, but by a

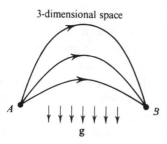

Figure 26. Particles do not move along geodesics in position space.

variety of paths (figure 26), depending on the initial velocity, for which the final events are different because they occur at different times. It is the *world lines* in *spacetime* that are unique, and take the form of geodesics.

4.3 Curved spaces

We introduced geodesics during a discussion of the straight world line of a free body in special relativity. Generally, however, the geodesic world lines of bodies under the action of gravity are *curved*, and there exists no coordinate transformation that will make them look straight, except in one special case. This occurs when the gravitational field **g** always points in the same direction, that is, when the 'lines of force' are parallel; an example is the Earth's gravitational field near a small region of the Earth's surface, where **g** = const. = 9.81 m s^{-2}, directed downwards. In this case the familiar parabolic paths that we see from the Earth's surface look straight if we transform to a freely-falling frame. Generally this is not possible, because only *locally* inertial frames exist (we ignore the trivial case of the single local inertial frame falling with the body at its spatial origin). Thus we now have geodesics – shortest or longest lines – that are intrinsically curved. This suggests that *spacetime itself is curved.*

It is difficult enough to imagine a curved three-dimensional space, let alone a curved four-dimensional spacetime (with indefinite metric to boot), so we explain the idea of curvature by considering *two-dimensional surfaces*, with which we are all familiar. Consider the following surfaces: a plane, a sphere and a cylinder; are these surfaces curved or flat? Obviously the plane is flat. The sphere is curved in a fundamental way: it cannot be deformed to coincide with the plane without stretching or tearing. The curvature of the cylinder is less fundamental: it can be simply unrolled onto the plane without distortion. We arrived at these conclusions by considering the three surfaces as *embedded* in three-dimensional position space. It was the great achievement of Gauss in the early nineteenth century to

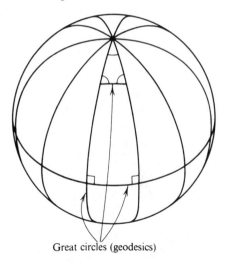

Great circles (geodesics)

Figure 27. Angle sum of spherical triangle exceeds 180°.

show that the curvature, and indeed the whole geometry, of a surface can be determined *intrinsically*, that is by measurements made in the surface itself by imaginary two-dimensional beings with measuring tapes.

Geometries are mathematical theories; to test which geometry applies to a given surface, we need physical definitions of the quantities appearing in the theory. We define the distance along any line as the number of units on a standard measuring tape coincident with the line, and we define the geodesic between two points on the surface as the line formed by a string in the surface stretched taut between the points. Then, if any of the familiar theorems of Euclidean plane geometry is found by experiment to be false, the surface is curved, and its metrical relations – its geometry – must be *non-Euclidean*. For example, in a plane the angles of a triangle whose sides are straight lines (geodesics) add up to 180°. On the surface of a sphere, however, the angles of a 'geodesic triangle' always exceed 180° (figure 27), and may be as great as 900° (for small triangles the angle sum exceeds 180° only slightly). Therefore the surface of a sphere can be shown to be curved by purely intrinsic measurements. (In an attempt to test the Euclidicity of three-dimensional position space, Gauss employed surveying techniques to measure the angle sum of a triangle formed by three mountains, using light rays to define geodesics in space. He failed to detect any departure from 180° because his experiments were not accurate enough, and the same is true of ours today.)

To introduce the precise definition of curvature we follow Rindler (see the bibliography) and use a different theorem of plane geometry; the

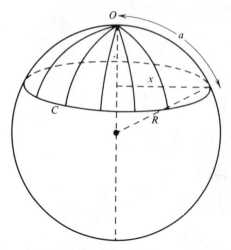

Figure 28. Geodesic circle of radius a on sphere of radius R.

circumference of a circle of radius a is $2\pi a$. To define a circle of radius a and centre O on an arbitrary surface, we draw all the geodesics emanating from O, and mark the point on each whose distance from O is a; the locus of all these points is the required circle. Let us try this out on a sphere of radius R (figure 28), and compute the circumference C. This is

$$C = 2\pi x = 2\pi R \sin \frac{a}{R},$$

▶ $$C = 2\pi a \left(1 - \frac{a^2}{6R^2} + \ldots\right).$$ (4.3.1)

If we define the curvature K of the sphere as

▶ $$K \equiv \frac{1}{R^2},$$ (4.3.2)

then the formula for C can be rearranged, to give

▶ $$K = \frac{3}{\pi} \lim_{a \to 0} \left(\frac{2\pi a - C}{a^3}\right).$$ (4.3.3)

This is a very useful result; it tells us that we can obtain the curvature of a sphere by measuring how the circumference of an infinitesimal circle departs from $2\pi a$. Exactly the same procedure can be employed to define the curvature of an arbitrary surface, whose curvature may vary from point to point. Also, curvature may be *negative*; this means that C exceeds $2\pi a$, and the corresponding surface looks saddle-shaped when embedded in a three-dimensional space. This is illustrated by figure 29, which shows

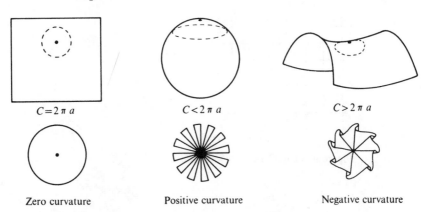

| $C=2\pi a$ | $C<2\pi a$ | $C>2\pi a$ |
| Zero curvature | Positive curvature | Negative curvature |

Figure 29. Circumference $\pm 2\pi \times$ radius on curved surfaces.

three surfaces (upper row), and what happens if we try to squash these surfaces onto a plane (lower row).

The analytical, as opposed to the geometrical, description of curvature employs the metric tensor $g_{\mu\nu}$. This gives the distance Δs between any two neighbouring points on the surface, in terms of the coordinate differences $\Delta x^\mu (\mu = 1, 2)$, where x^μ are coordinates on an arbitrarily-drawn grid lying in the surface. Δs is given by equation (4.1.7). In a plane, if we use a Cartesian grid where $x^1 = x$, $x^2 = y$ (figure 30), then

$$\Delta s^2 = \Delta x^2 + \Delta y^2 = (\Delta x^1)^2 + (\Delta x^2)^2,$$

and $\quad g_{\mu\nu} = \begin{pmatrix} 1 & 0 \\ 0 & 1 \end{pmatrix}.$

Pythagoras' theorem is satisfied, indicating that we are on a flat surface. However, we could equally well use polar coordinates $x^1 = r$, $x^2 = \phi$ (figure 31) and then

$$\Delta s^2 = \Delta r^2 + r^2 \Delta\phi^2 = (\Delta x^1)^2 + (x^1)^2 (\Delta x^2)^2,$$

and

$$g_{\mu\nu} = \begin{pmatrix} 1 & 0 \\ 0 & (x^1)^2 \end{pmatrix}.$$

Now the metric tensor is position-dependent, yet we are still dealing with the same flat surface. If we were simply given the metric distance formula in polar coordinates, how would we know whether the surface is flat or not? By seeking a coordinate transformation which would give us back the

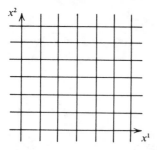

Figure 30. Plane Cartesian coordinates.

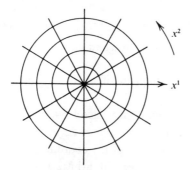

Figure 31. Plane polar coordinates.

Cartesian distance formula. In this case the new coordinates would be $x^{1'}$ and $x^{2'}$, where

$$x^{1'} = x^1 \cos x^2, \quad \text{i.e.} \quad x = r \cos \phi$$
$$x^{2'} = x^1 \sin x^2, \quad \text{i.e.} \quad y = r \sin \phi.$$

Of course, $x^{1'}$ and $x^{2'}$ are just our old x and y.

Now let us play this game with a cylinder of radius R. With cylindrical coordinates z, r, ϕ in the three-dimensional embedding space (figure 32), the surface is defined by $r = \text{const.} = R$, and the distance formula is

$$\Delta s^2 = \Delta z^2 + R^2 \Delta \phi^2.$$

It is easy to show that this is flat: define coordinates

$$x^1 = z, x^2 = R\phi.$$

Then, since R is constant,

$$\Delta s^2 = (\Delta x^1)^2 + (\Delta x^2)^2,$$

which is the same as the distance formula on a plane. Therefore the surface of a cylinder is intrinsically flat. (We are not saying that a cylinder is the same as a plane in all respects. Clearly it is not: of all the geodesics through any point on a cylinder, one will be closed (figure 33); this does not happen

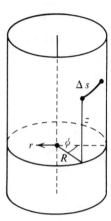

Figure 32. Line element on cylinder.

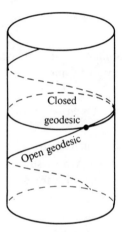

Figure 33. Closed and open geodesics on cylinder.

on a plane. Differences of this sort are *global*, that is, differences in the connectivity of the surface as a whole, or *topological* differences. Flatness, on the other hand, is a *local* property, and can be established by measuring small distances, triangles, circles etc.)

Now we consider the surface of a sphere of radius R. This time convenient coordinates x^1 and x^2 are the two angles θ and ϕ of spherical polar coordinates. On the earth, $\pi/2 - \theta$ corresponds to latitude, and ϕ to longitude (figure 34). The distance formula is

$$\Delta s^2 = R^2 \Delta\theta^2 + R^2 \sin^2\theta \, \Delta\phi^2$$
$$= R^2(\Delta x^1)^2 + R^2 \sin^2 x^1(\Delta x^2)^2,$$

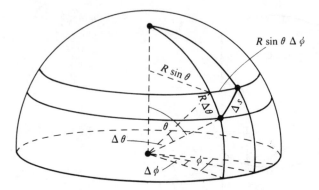

Figure 34. Line element on sphere.

and the metric tensor is

$$\blacktriangleright \qquad g_{\mu\nu} = \begin{pmatrix} R^2 & 0 \\ 0 & R^2 \sin^2 x^1 \end{pmatrix}. \qquad (4.3.4)$$

Now we test for flatness by seeking new coordinates $x^{1'}$ and $x^{2'}$, which are functions of x^1 and x^2 (i.e. of θ and ϕ); in terms of $x^{1'}$ and $x^{2'}$ the distance must take the Cartesian form

$$(\Delta x^{1'})^2 + (\Delta x^{2'})^2.$$

But however hard we try, we cannot find such a transformation, and it appears that the metrical properties of a spherical surface are intrinsically different from those of a plane, and the form for Δs^2 does not result simply from an inappropriate choice of coordinates.

However, this is certainly not a proof; how can we know from the metric tensor alone that there may not be some obscure transformation which will do the trick, and reduce a spherical surface to a plane? The answer was given by Gauss, who calculated the circumference of a circle of radius a, and hence obtained the curvature K, in an arbitrary coordinate system in terms of the components $g_{\mu\nu}$ of the metric tensor. Before writing down Gauss's curvature formula, we make two remarks: firstly, if the $g_{\mu\nu}$ are all constants, it is trivial to find a transformation to $\Delta s^2 = \Delta x^2 + \Delta y^2$; therefore it is the spatial variation of $g_{\mu\nu}$ that determines curvature, and we expect Gauss's formula to involve *derivatives* of $g_{\mu\nu}$ with respect to x^1 and x^2. Secondly, any $g_{\mu\nu}$ can always be transformed to *diagonal form*, where $g_{12} = g_{21} = 0$; the metric is then called *orthogonal*, since the grid lines for x^1 and x^2 always cross at right angles; all the metrics we shall deal with in relativity and cosmology will be orthogonal, so we write Gauss's formula for this particular case, where only g_{11} and g_{22} are involved. Now for the formula: it is

▶ $$K = \frac{1}{2g_{11}g_{22}} \left\{ -\frac{\partial^2 g_{11}}{\partial (x^2)^2} - \frac{\partial^2 g_{22}}{\partial (x^1)^2} + \frac{1}{2g_{11}} \left[\frac{\partial g_{11}}{\partial x^1} \frac{\partial g_{22}}{\partial x^1} + \left(\frac{\partial g_{11}}{\partial x^2} \right)^2 \right] \right.$$
$$\left. + \frac{1}{2g_{22}} \left[\frac{\partial g_{11}}{\partial x^2} \frac{\partial g_{22}}{\partial x^2} + \left(\frac{\partial g_{22}}{\partial x^1} \right)^2 \right] \right\}. \tag{4.3.5}$$

The proof is given in Appendix B. Obviously K is zero when $g_{\mu\nu}$ corresponds to a plane with a Cartesian grid, because the $g_{\mu\nu}$ are constants. Calculation also confirms that K vanishes for a plane with polar coordinates, where g_{22} depends on x^1 so that the derivatives are not zero. This suggests that K as given by (4.3.5) is *invariant*, that is, no matter what coordinate system we use, we always get the same value for K, which is therefore an intrinsic geometrical property of the surface itself. Gauss was able to prove this and to show, moreover, that K is the *only* invariant of the surface that can be constructed from derivatives of $g_{\mu\nu}$ no higher than the second.

Now we return to the case of the spherical surface. The metric tensor is given by (4.3.4), and the curvature formula (4.3.5) gives $K = 1/R^2$, as we would expect from (4.3.2). Thus we know definitely that the sphere is intrinsically curved. Note, by the way, that it is perfectly possible to have surfaces whose curvature varies from point to point – the invariance of K means that its value *at any point* is coordinate-independent, not that its value is the same at all points.

When considering the curvature of spaces whose dimensionality exceeds two, our intuition fails, because the 'embedding spaces' have dimensionality greater than three, and we cannot imagine them. It is in these cases that the intrinsic methods of investigating curvature come into their own. We can, for example, measure the angles of a geodesic triangle in three-dimensional space; if their sum differs from 180°, the space is curved. By filling the area between the sides of the triangle with geodesics, we define a surface. A point on that surface may be said to be 'on' the triangle. But it also lies on an infinite number of other triangles whose surfaces intersect the first (figure 35). This means that in spaces of more than two dimensions it is not possible to specify curvature by just one function K. It turns out that a fourth rank 'curvature tensor' R_{ijkl} is necessary. This is defined in terms of derivatives of the metric tensor, just like K. R_{ijkl} has n^4 components in n dimensions, but these are not all independent, because many symmetry relations exist (e.g. $R_{ijkl} = R_{klij}$). In two dimensions there are just fifteen such relations, so that only one independent component

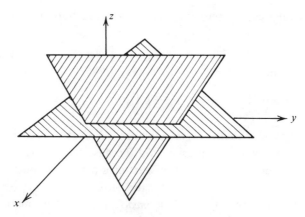

Figure 35. Two geodesic surfaces through the same point.

($= 2^4 - 15$) exists; this is K, of course. Fortunately for our problems in relativity and cosmology we shall not need this general theory, because we shall deal with spaces and spacetimes with special symmetries.

One such symmetrical case is the three-dimensional isotropic space with constant curvature K. Let us work out its metric tensor $g_{\mu\nu}$, and also the area of 'hyperspheres' in this space; the results will be useful in chapter 6 when we develop relativistic cosmology. We shall employ polar coordinates r, θ, ϕ, resembling those in Euclidean space. The radial coordinate r defines a 'hyperspherical' surface of rigid rods, whose area is *by definition* $4\pi r^2$ (in the case of a two-dimensional curved surface the coordinate analogous to r is the x in figure 28). On this hyperspherical surface r is constant and positions are specified by θ and ϕ. The metric of this two-dimensional subspace is given by the 'sphere' formula (4.3.4) as

$$\Delta s^2_{(r=\text{const.})} = r^2\,\Delta\theta^2 + r^2\sin^2\theta\,\Delta\phi^2.$$

A succession of concentric 'r-spheres', each made of rigid rods, defines coordinates throughout the three-dimensional space. However – and this is very important – r is *not* the proper radius of each sphere, because the space is curved. If we stretched a measuring tape between the origin and the surface of the r-sphere, its length would not be r. To allow for this effect, we write the metric as

▶ $$\Delta s^2 = f(r)\,\Delta r^2 + r^2\Delta\theta^2 + r^2\sin^2\theta\,\Delta\phi^2, \qquad (4.3.6)$$

where $(\sqrt{f(r)})\,\Delta r$ is the proper distance between neighbouring points (r, θ, ϕ) and $(r+\Delta r, \theta, \phi)$ on the same radius, and where the last two terms express the isotropy of the space.

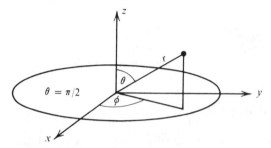

Figure 36. 'Equatorial' surface.

To find $f(r)$ we realise that in our symmetrical space all geodesic surfaces must have the same curvature K. We choose the 'equatorial' surface $\theta = \pi/2$ (figure 36). Then $\Delta\theta = 0$ and

$$\blacktriangleright \quad \Delta s^2_{(\theta=\pi/2)} = f(r)\,\Delta r^2 + r^2\Delta\phi^2. \qquad (4.3.7)$$

In this surface we have coordinates $x^1 = r$, $x^2 = \phi$, and a metric

$$g_{\mu\nu} = \begin{pmatrix} f(x^1) & 0 \\ 0 & (x^1)^2 \end{pmatrix}, \quad \text{i.e.} \quad g_{11} = f(x^1),\ g_{22} = (x^1)^2.$$

The curvature can be calculated by Gauss's formula (4.3.5), which gives

$$K = \frac{df(x^1)/dx^1}{2f^2(x^1)\,x^1}.$$

Because K is constant, this differential equation for the unknown function $f(x^1)$ can easily be solved, as follows:

$$\frac{df(x^1)/dx^1}{f^2(x^1)} = -\frac{d}{dx^1}\left(\frac{1}{f(x^1)}\right) = 2Kx^1,$$

Therefore

$$1/f(x^1) = C - K(x^1)^2,$$

i.e. $\quad f(x^1) = 1/(C - K(x^1)^2).$

C is a constant of integration, which we can determine from the boundary condition that, for a flat space $(K = 0)$, $r(= x^1)$ should reduce to an ordinary radial proper distance coordinate Thus $f = 1$ if $K = 0$, so that $C = 1$. The final metric in this constant-curvature space is therefore

$$\blacktriangleright \quad \Delta s^2 = \Delta r^2/(1 - Kr^2) + r^2\Delta\theta^2 + r^2\sin^2\theta\,\Delta\phi^2. \qquad (4.3.8)$$

The area A of the r-sphere is $4\pi r^2$ by definition, and the proper radius is $a(r)$, given by

$$a(r) = \int_0^{a(r)} ds = \int_0^r \frac{dr}{\sqrt{(1 - Kr^2)}} = \frac{1}{\sqrt{K}}\arcsin\left(r\sqrt{K}\right),$$

so that

$$r = \frac{1}{\sqrt{K}} \sin{(a\sqrt{K})}$$

and the relationship between area and proper radius for these hyperspheres is

▶ $$A = \frac{4\pi}{K} \sin^2{(a\sqrt{K})}. \tag{4.3.9}$$

For small spheres ($a \ll 1/\sqrt{K}$), the area A approximates to the Euclidean value $4\pi a^2$. As a increases, A departs from the Euclidean value in a way which depends on the sign of K. If K is negative, we can write (4.3.9) in the form

▶ $$A = \frac{4\pi}{|K|} \sinh^2{(a\sqrt{|K|})} \quad (K = -|K|), \tag{4.3.10}$$

and this shows that the area increases faster than in flat space, and tends to infinity with the radius. If the space is positively curved, however, (4.3.9) shows that A increases more slowly than $4\pi a^2$, and reaches a maximum value

▶ $$A_{max} = \frac{4\pi}{K} \quad (K > 0, a = \pi/2\sqrt{K}). \tag{4.3.11}$$

As the proper radius a increases further, A decreases and becomes zero when $a = \pi/\sqrt{K}$ This means that this positively-curved space is *closed*, and the periodic behaviour of A as given by formula (4.3.9) corresponds to successive circumnavigations. The situation is closely analogous to that described by (4.3.1), namely the behaviour of the *circumference* of *circles* of different proper radius a on a *surface* of constant curvature $1/R^2$ (see figure 28); if the origin O of the space corresponds to the north pole of the spherical surface, then the surface of the hypersphere of proper radius $a = \pi/2\sqrt{K}$ corresponds to the equatorial circle, and the zero-area hypersphere of proper radius $a = \pi/\sqrt{K}$ corresponds to the south pole on the surface.

4.4 Curvature and gravitation
Spacetime is curved near gravitating matter. To see this, consider the separation formula (4.1.3) of special relativity. This spacetime is flat, because the components g_{ij} of the metric tensor are constants. We know, however, that this formula applies only in an inertial frame, and it is impossible to cover spacetime with one such frame if there is a non-homogeneous gravitational field present. Only *local* inertial frames exist, so that

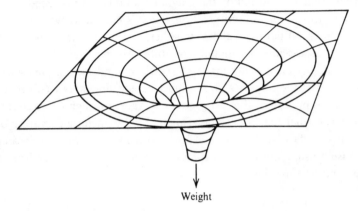

Weight

Figure 37. Rubber sheet analogy for spacetime curved by matter.

spacetime appears flat on a small scale. This statement is precisely analogous to the statement that all surfaces are locally flat ($C \to 2\pi a$ as $a \to 0$ etc.). In the large, however, there is no frame of reference – no coordinate system – in which the separation takes the form (4.1.3) of special relativity. Therefore spacetime is not flat, in general.

Mass 'warps' spacetime in its vicinity, rather as a weight curves the surface of a rubber sheet (figure 37). The world lines of planets, etc., are geodesics in this curved spacetime. The precise relationship between gravitating matter and the curvature of space is expressed by *Einstein's field equations*. These state that a certain tensor describing the distribution of matter (which may be continuous, or in the form of discrete masses) equals a tensor describing the curvature of spacetime. Both of these tensors are essentially unique, so that the field equations form a very natural description of the gravitational effects of matter, they are also generally covariant. Finally, they solve the problem posed in section 4.2 of what determines the metric tensor g_{ij}; it is determined by solving the field equations in the presence of a given distribution of matter, since the curvature tensor involves g_{ij}.

Although the field equations, and the arguments leading to their postulation, are almost unparalleled in their formal beauty, the full generality is unnecessary for our purposes. Therefore we present now an argument leading to the connection between matter and curvature in a simple case: we consider the neighbourhood of a point mass M (this is the active gravitational mass). Spacetime will be curved, and by employing a simple 'clock-and-rod' coordinatisation we can see that the curvature has two aspects: time is distorted by the gravitational dilation discussed in section 3.6, and three-dimensional position space is curved as well. In the next section we shall put these effects together, and derive the whole tensor g_{ij}; here we discuss just the spatial curvature. Again we employ coordinates r, θ, ϕ, the r-sphere being made of rigid rods and having area $4\pi r^2$. The mass M is at the origin O. Space must be isotropic about O, because no direction is preferred; thus all geodesic surfaces through O are equivalent. This means that the space is determined by the intrinsic curvature of any one such surface. This will be a function $K(r)$ of the radial coordinate r, because we shall not have constant curvature in this case. Far from O the space must be flat, i.e.

$$K(r) \to 0 \quad \text{as} \quad r \to \infty.$$

Now we ask: what is the simplest possible form for $K(r)$? Since the curvature is produced by the mass M, we assume that K is proportional to M. Since K must vanish for large r, we assume a power law decay, that is, K is proportional to r^{-n}, where n is to be determined. Now K has dimensions $(\text{length})^{-2}$, and no combination of the form Mr^{-n} has these dimensions. However, we expect the gravitational constant G to figure in any formula for K, and also possibly the speed of light c. Therefore we postulate

$$K(r) = qMG^l c^m r^{-n}$$

as the curvature law, q being a dimensionless constant and l, m and n indices to be determined. On dimensional grounds, the only possible values for l, m and n are 1, -2 and 3 respectively. This leaves q; is it positive or negative? Let us take the rubber-sheet analogy seriously. The sheet is negatively curved by the weight (figure 37). Therefore the simplest choice of q is -1. Thus our postulated 'simplest possible' gravity law is:

▶ $$K(r) = -GM/c^2 r^3. \tag{4.4.1}$$

More complicated laws can be imagined, for instance

$$K(r) = -(GM/c^2)^2 \exp\left(-rc^2/GM\right)/r^4.$$

However our law (4.4.1) is the one which emerges rigorously from Einstein's field equations (this is not surprising, because these laws are also the 'simplest possible'). The argument leading to the choice $q = -1$ is unconvincing, but we shall be using the result rather soon; a more logical approach would be to fix q by demanding that dynamical equations be correct in the Newtonian limit (section 5.3).

The curvature of space predicted by (4.4.1) is usually very small in comparison with the curvature of the r-sphere, as the following values show: At the surface of the Sun,

$$\left|\frac{K(R_\odot)}{1/R_\odot^2}\right| = \frac{M_\odot G}{c^2 R_\odot} = 2.12 \times 10^{-6}.$$

At the surface of the Earth,

$$\left|\frac{K(R_\oplus)}{1/R_\oplus^2}\right| = \frac{M_\oplus G}{c^2 R_\oplus} = 6.97 \times 10^{-10}.$$

In view of the second result, it is not entirely surprising that Gauss's observations on a triangle formed by three mountains failed to detect the non-Euclidicity of space.

We must emphasise now that the result (4.4.1) describes a special case: the static spatial curvature near an isolated mass. In cosmological applications, where we have a continuous distribution of moving matter, the formula does not apply; however, it will be shown in section 7.1 that arguments similar to those used here can be employed to give the curvature of cosmologically interesting spacetimes. Nor does (4.4.1) apply to gravitational waves. These are 'ripples' in spacetime – that is, oscillatory distortions of the g_{ij} which are anything but isotropic and static. The simple treatment of section 3.4, based on the electromagnetic analogue, made the approximation of treating gravitation as a *vector field* (the acceleration **g**), rather than a *tensor field* (the metric g_{ij}).

The connection between the metric tensor g_{ij} and the distribution of matter completes the logical structure of general relativity. Knowing g_{ij} geodesics can be calculated, and the paths of particles and light rays predicted and compared with experiment. In chapter 5 we describe how general relativity has been tested to date; this will clear the way for cosmological applications of relativity, which begin in chapter 6.

5 General relativity near massive objects

5.1 Spacetime near an isolated mass

Near a body of mass M (e.g. the Sun), spacetime is curved. The world lines of particles and light rays in the field of M are geodesics; to find these it is necessary to know the metric tensor g_{ij} in some useful coordinate system x^i. We choose x^0 to be a time variable t, and x^1, x^2, x^3 to be the polar coordinates already introduced, with M at the origin O. Thus we have a series of concentric r-spheres with defined area $4\pi r^2$, each sphere being made of rigid rods arranged in a grid corresponding to latitude and longitude angles θ and ϕ, with clocks at each intersection. If M were zero, the separation formula would be

$$\Delta\tau^2 = \Delta t^2 - (\Delta r^2 + r^2\Delta\theta^2 + r^2\sin^2\theta\,\Delta\phi^2)/c^2,$$

as in special relativity (cf. (4.1.3)). If we now 'switch on' the mass M, two things happen: the position space becomes curved, so that the r-spheres no longer lie at proper distance r from O, and clocks on each r-sphere are no longer observed from other r-spheres to run at the same rate. To allow for these effects we write the separation as

▶ $$\Delta\tau^2 = e(r)\,\Delta t^2 - (f(r)\,\Delta r^2 + r^2\Delta\theta^2 + r^2\sin^2\theta\,\Delta\phi^2)/c^2; \qquad (5.1.1)$$

$e(r)$ and $f(r)$ are functions to be determined, subject to the boundary condition that far from M spacetime is flat, so that

$$e(\infty) = f(\infty) = 1.$$

By solving Einstein's field equations the function $e(r)$ can be found to be

$$e(r) = 1 - 2GM/c^2r.$$

We shall not be able to prove this result rigorously using our simplified methods, but we can make it plausible by the following discussion of *time measurements* near the mass M. Let an observer O_2 with a clock C_2 at (r_2, θ, ϕ) (figure 38) look at (i.e. receive light from) an identical clock C_1 at $(r_1(< r_2), \theta, \phi)$. (For example, each tick of C_1 might be accompanied by a

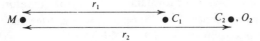

Figure 38. Clocks in Schwarzschild spacetime.

flash of light, or C_1 might be an atomic clock, emitting radiation whose wave crests correspond to ticks.) O_2 will not see C_1 reading the same time as his own clock C_2, and he will not expect to, because he knows that the light takes time to travel from C_1 to him. But he does not even see C_1 ticking at the same rate as C_2, because of the gravitational time dilation effect. In fact he sees C_1 run more slowly, and when C_2 indicates that one hour has elapsed, we see from equation (3.6.4) that O_2 observes C_1 (at a distance) indicating the passage of only

$$[1 - (GM/c^2r_1 - GM/c^2r_2)] \text{ hours.}$$

Thus O_2 will see his own clock getting more and more out of step with C_1. This is an inconvenient property for a time coordinate; we require the outer observer to see the other's clock always lagging behind his own by the same amount, which corresponds to the time for light to travel to him. Only then will the spacetime be static in the sense that its metric formula is independent of t.

To achieve this, we speed up all the coordinate clocks on the r-sphere by the Doppler factor

$$[1 - (GM/c^2r)]^{-1}.$$

Then all the clocks appear to run at the same rate as the clocks at infinity, which run undisturbed at their proper rates; moreover, any observer will see all the other clocks running at the same rate as his own. This result is general, and holds for all pairs of clocks, even those which do not have the same θ and ϕ coordinates. The speeding-up of clocks means that an indicated time difference Δt corresponds to a smaller proper time $\Delta \tau$ for the clock, namely

$$\Delta \tau = (1 - GM/c^2r) \Delta t, \quad \text{i.e. } \Delta \tau^2 = (1 - 2GM/c^2r) \Delta t^2,$$

where in the second equation we have neglected a term $(GM/c^2r)^2$. On the world line of a clock, $\Delta r = \Delta \theta = \Delta \phi = 0$, so that we can also obtain $\Delta \tau^2$ from the separation formula (5.1.1), and identify $e(r)$ as

$$e(r) = 1 - 2GM/c^2r.$$

This is identical with the exact result previously stated; what has happened is that our neglect of $(GM/c^2r)^2$ cancelled a previous approximation introduced by using the Newtonian 'work' formula in the original derivation of the time dilation formula (3.6.4).

To find $f(r)$ we employ Gauss's formula (4.3.5) for the curvature $K(r)$ of the two-dimensional subspace $\Delta\theta = \Delta t = 0$, $\theta = \pi/2$, together with the 'simplest possible' gravity formula (4.4.1). This gives

$$K(r) = \frac{df(r)}{dr} \Big/ 2rf^2(r) = -GM/c^2r^3.$$

Integration leads to

$$-1/f = 2GM/c^2r + \text{const.} = 2GM/c^2r - 1,$$

where the constant has been determined by the condition $f(\infty) = 1$. Thus

$$f(r) = \frac{1}{1-2GM/c^2r}.$$

We can now write down the complete separation formula:

▶ $$\Delta\tau^2 = (1 - 2GM/c^2r)\,\Delta t^2$$
$$-\frac{1}{c^2}\left(\frac{\Delta r^2}{1-2GM/c^2r} + r^2\,\Delta\theta^2 + r^2\sin^2\theta\,\Delta\phi^2\right). \qquad (5.1.2)$$

This is known as the 'Schwarzschild metric', after Schwarzschild, who derived it rigorously from Einstein's field equations in 1916. The corresponding metric tensor is

▶ $$g_{ij} = \begin{pmatrix} 1-2GM/c^2r & 0 & 0 & 0 \\ 0 & -\dfrac{1/c^2}{1-2GM/c^2r} & 0 & 0 \\ 0 & 0 & -r^2/c^2 & 0 \\ 0 & 0 & 0 & -\dfrac{r^2\sin^2\theta}{c^2} \end{pmatrix}.$$

$$(5.1.3)$$

These relationships enable the behaviour of test bodies, light rays and clocks to be predicted, so that the general theory of relativity can be tested. We shall describe some of these tests in the next few sections. The mass M will refer either to the Sun or the Earth or (in section 5.6) to massive stars.

5.2 Around the world with clocks

Consider two identical clocks A and B, initially synchronised in setting and rate and at rest on the Earth's surface. A remains at rest, while B is flown round the world at height h in an aircraft whose speed relative to the ground is v (figure 39). After the circumnavigation, A and B are compared, that is, the proper time τ_A experienced by A is compared with the proper time τ_B experienced by B. We do not expect τ_A and τ_B to be

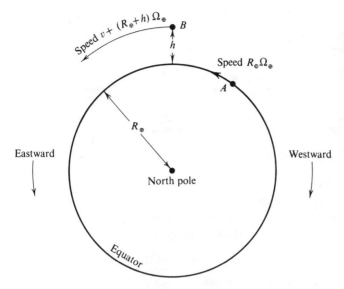

Figure 39. Clocks going round (*A*) on the ground, (*B*) in an aircraft.

the same, because *A* and *B* have pursued different world lines between the initial (synchronisation) event *I* and the final (comparison) event *F*.

To calculate τ_A and τ_B we use the Schwarzschild metric (5.1.2), assume for simplicity that the circumnavigation is equatorial, and remember that the earth is rotating with angular velocity $\Omega_\oplus$ relative to the local inertial frame in which it is embedded (i.e. a frame in free fall in the field of the Sun). For *A*, we have $r = R_\oplus$, $\Delta r = 0$, $\Delta\theta = 0$, $\theta = \pi/2$. Thus

$$\tau_A = \int \Delta\tau_A = \int\sqrt{[(1 - 2GM_\oplus/c^2 R_\oplus)\,\Delta t^2 - R_\oplus^2\,\Delta\phi^2/c^2]}.$$

Now the speed of *A* relative to the local inertial frame is

$$R_\oplus\Omega_\oplus = R_\oplus(d\phi/dt),$$

so that

$$\tau_A = \int_0^t dt\,\sqrt{[(1 - 2GM_\oplus/c^2 R_\oplus) - R_\oplus^2\,\Omega_\oplus^2/c^2]}$$
$$= t\sqrt{(1 - 2GM_\oplus/c^2 R_\oplus - R_\oplus^2\,\Omega_\oplus^2/c^2)}$$

where *t* is the coordinate time between the events *I* and *F*.

For clock *B* we neglect the small effects of rising to, and descending from, the height *h* (which can in any case be rendered as small as we like by repeated circumnavigation), and write

$$\tau_B = \int\sqrt{[1 - 2GM_\oplus/c^2(R_\oplus + h) - (R_\oplus + h)^2(\Delta\phi/\Delta t)^2/c^2]}\,\Delta t.$$

Table 2

Direction of circumnavigation	$\tau_B - \tau_A$ (nanoseconds)	
	Experiment	Theory
Westward	273 ± 7	275 ± 21
Eastward	-59 ± 10	-40 ± 23

The coordinate speed of B is

$$(R_\oplus + h)(\Delta\phi/\Delta t) \approx (R_\oplus + h)\,\Omega_\oplus + v,$$

where the equality is only approximate because we have neglected the relativistic correction to the velocity addition formula. A positive v refers to eastward motion (i.e. motion with the earth). Thus

$$\tau_B = t\sqrt{\{1 - 2GM_\oplus/c^2(R_\oplus + h) - [(R_\oplus + h)\,\Omega_\oplus + v]^2/c^2\}}$$

where t is again the coordinate time between I and F.

Now we can calculate the 'time offset' δ, defined as

$$\delta \equiv (\tau_B - \tau_A)/\tau_A.$$

The coordinate time t cancels, and we can, moreover, simplify the square roots, because

$$GM_\oplus/c^2 R_\oplus \ll 1,\ h/R_\oplus \ll 1,\ v^2/c^2 \ll 1,\ h\Omega_\oplus/v \ll 1.$$

To the lowest order in these four small quantities, we get

▶ $$\delta = gh/c^2 - (2R_\oplus \Omega_\oplus + v)\,v/2c^2, \tag{5.2.1}$$

where

$$g = GM_\oplus/R_\oplus^2 = 9.81 \text{ m s}^{-2}.$$

The time offset is very small: if we take $h = 10^4$ m and a typical jet speed $v = 300$ m s^{-1}, we obtain

$$\delta_{\text{westward}} = 2.1 \times 10^{-12} \quad \text{and} \quad \delta_{\text{eastward}} = -1.0 \times 10^{-12}.$$
$$\underset{(v<0)}{} \qquad\qquad\qquad\qquad \underset{(v>0)}{}$$

Nevertheless, the fractional accuracy of modern caesium clocks is about 10^{-13}, so that the effect is observable. In 1971 Hafele and Keating made eastward and westward journeys round the world on commercial jet flights, carrying caesium clocks which they later compared with clocks which remained at the US Naval Observatory in Washington. They did not quite employ formula (5.2.1) for δ, because the flights were not equatorial. However, the theory is essentially the same, and the results, shown in table 2, represent a successful direct test of the dependence of the separation between two events on the world line joining them, and conclusively resolve the 'twin paradox' controversy in Einstein's favour. It also successfully tests the Schwarzschild metric formula (5.1.2).

5.3 Precession of the perihelion of Mercury

In this section we calculate the centennial precession $\Delta\phi^{100}$ of planetary orbits; this was introduced in section 3.2 as a deviation from the predictions of Newtonian mechanics. The motion of a planet according to general relativity is a timelike geodesic in the Schwarzschild spacetime surrounding the Sun. Because of the spherical spatial symmetry, the path of the planet remains in the plane containing the initial direction and the Sun, just as in Newtonian mechanics. Therefore we lose no generality by restricting the motion at the outset to the equatorial plane $\theta = \pi/2$. The world line is then specified by the three functions $t(\tau)$, $r(\tau)$ and $\phi(\tau)$, which are determined (cf. section 4.2) by the geodesic condition

$$\blacktriangleright \quad \tau_{AB} = \int_{\tau_A}^{\tau_B} d\tau \sqrt{\left(g_{ij}\frac{dx^i}{d\tau}\frac{dx^j}{d\tau}\right)}$$

$$= \int_{\tau_A}^{\tau_B} d\tau \sqrt{\left[\left(1-\frac{2GM}{c^2r}\right)t'^2 - \frac{1}{c^2}\left(\frac{r'^2}{1-2GM/c^2r}+r^2\phi'^2\right)\right]}$$

$$= \text{extremum}, \tag{5.3.1}$$

where A and B are any two events on the planet's world line, and t', r' and ϕ' stand for $dt/d\tau$, $dr/d\tau$ and $d\phi/d\tau$.

We could get three equations from this condition, and thus determine $t(\tau)$, $r(\tau)$ and $\phi(\tau)$; however, we need to derive only two since we know from the basic metric formula (4.1.9) that

$$\blacktriangleright \quad g_{ij}\frac{dx^i}{d\tau}\frac{dx^j}{d\tau} = \left(1-\frac{2GM}{c^2r(\tau)}\right)[t'(\tau)]^2$$

$$-\frac{1}{c^2}\left[\frac{(r'(\tau))^2}{1-2GM/c^2r(\tau)}+(r(\tau)\,\phi'(\tau))^2\right] = 1. \tag{5.3.2}$$

To get 'equations of motion' from the geodesic condition (5.3.1) we imagine the planet to have a world line W between A and B slightly different from the actual geodesic world line. Along this 'varied' world line the separation is τ_{AB}^W, and the extremum condition implies that

$$\delta\tau_{AB} \equiv \tau_{AB}^W - \tau_{AB} = 0.$$

First we take for W a time-varied world line (figure 40); that is, we assume that the functions $r(\tau)$ and $\phi(\tau)$ describing the planet's path in space are fixed, but that the time $t(\tau)$ when the planet reaches different parts of its

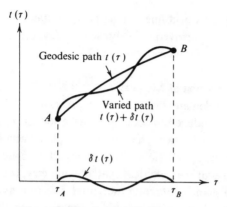

Figure 40. Geodesic and 'varied' path between events.

orbit is altered in such a way that

$$t(\tau) \rightarrow t(\tau) + \delta t(\tau)$$

where $\delta t(\tau)$ is any small function of τ which vanishes at A and B. It is convenient to denote by $f(\tau, t'(\tau))$ the square root of $g_{ij}(\mathrm{d}x^i/\mathrm{d}\tau)(\mathrm{d}x^i/\mathrm{d}\tau)$ that enters into (5.3.1). Then we have

$$\delta\tau_{AB} = \int_{\tau_A}^{\tau_B} \mathrm{d}\tau[f(\tau, t'(\tau) + \delta t'(\tau)) - f(\tau, t'(\tau))]$$

$$= \int_{\tau_A}^{\tau_B} \mathrm{d}\tau\left[\frac{\partial f(\tau, t'(\tau))}{\partial t'}\right]\delta t'(\tau),$$

where we have used the fact that $\delta t(\tau)$ is small (strictly, it is an infinitesimal of the first order). Now we integrate by parts; the 'integrated' term is zero (why?) and we obtain

$$\delta\tau_{AB} = -\int_{\tau_A}^{\tau_B} \mathrm{d}\tau\left[\frac{\mathrm{d}}{\mathrm{d}\tau} \cdot \frac{\partial f(\tau, t'(\tau))}{\partial t'}\right]\delta t(\tau) = 0.$$

Now, $\delta t(\tau)$ is an *arbitrary* small function, so the integral can vanish only if

$$\frac{\mathrm{d}}{\mathrm{d}\tau}\frac{\partial f(\tau, t'(\tau))}{\partial t'} = 0, \quad \text{i.e.} \quad \frac{\partial f(\tau, t'(\tau))}{\partial t'} = \text{const.}$$

Using the explicit form for f, and the fact that its value is unity (equation (5.3.2)), we arrive finally at the equation of motion for $t(\tau)$, namely

▶ $$(1 - 2GM/c^2r(\tau))\, t'(\tau) = \text{const.} \equiv A. \tag{5.3.3}$$

By a precisely analogous argument we can 'vary' the function $\phi(\tau)$, keeping $t(\tau)$ and $r(\tau)$ fixed; this procedure gives the equation of motion for $\phi(\tau)$, namely

▶ $$r^2(\tau)\,\phi'(\tau) = \text{const.} \equiv B. \tag{5.3.4}$$

Thus we have three equations of motion (5.3.2), (5.3.3) and (5.3.4), and we can find the world line of a planet.

At this point we can make connections with the analogous equations of Newtonian gravitational theory. In Newtonian mechanics, r is an ordinary radial coordinate, and there is no distinction between t and τ. With these identifications, the ϕ equation (5.3.4) simply expresses the *conservation of angular momentum*, the value of this quantity being mB, where m is the mass of the planet. To see this, we write mB as

$$mB = mr^2\phi' = m \times r \times r\phi' = \text{mass} \times \text{distance} \times \text{transverse velocity}.$$

To interpret (5.3.2), we combine it with (5.3.3) and rearrange; this gives

▶ $$\tfrac{1}{2}[\underbrace{(r'(\tau))^2}_{\substack{\text{(radial}\\ \text{velocity)}^2}} + \underbrace{(r(\tau)\,\phi'(\tau))^2}_{\substack{\text{(transverse}\\ \text{velocity)}^2}}\underbrace{(1-2GM/c^2r(\tau))}_{\substack{\text{relativistic}\\ \text{correction}}}] - \underbrace{GM/r(\tau)}_{\substack{\text{gravitational}\\ \text{potential}}}$$

$$= (A^2-1)\,c^2/2 \equiv \underbrace{\mathscr{E}}_{\substack{\text{total}\\ \text{energy}}\,\div\,m} \tag{5.3.5}$$

We recognise the *conservation of energy*, exact apart from the 'general relativity' term $2GM/c^2r$ which vanishes in the Newtonian limit $c \to \infty$. Thus the connection with Newtonian theory is very close, as of course it must be. The conceptual basis of general relativity, however, is completely different: instead of inertial, passive and active gravitational masses, forces, and accelerations, we have simply active gravitational mass, and geodesics in curved spacetime.

Now we return to the calculation of orbits. We are not interested in knowing the reading τ on the planet's clock at the event (t, r, ϕ) but simply in the *shape of the orbit in space*; thus we require the radius $r(\phi)$ as a function of the azimuth angle ϕ. We find $r(\phi)$ from (5.3.5), after eliminating $r'(\tau)$ with the aid of the relation

$$r'(\tau) = \frac{dr(\phi)}{d\phi}\,\phi'(\tau),$$

and eliminating $\phi'(\tau)$ with the aid of (5.3.4); this gives

▶ $$\frac{B^2}{2r^2(\phi)}\left[\frac{(dr(\phi)/d\phi)^2}{r^2(\phi)} + 1 - 2GM/c^2r(\phi)\right] - GM/r(\phi) = \mathscr{E}, \tag{5.3.6}$$

an equation determining the family of orbits $r(\phi)$ specified by fixing the values of the two constants $\mathscr{E}$ and B.

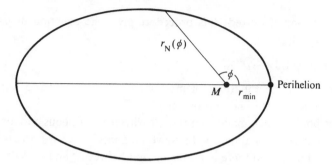

Figure 41. Notation for Newtonian orbit.

In Newtonian mechanics ($c \rightarrow \infty$) orbits are specified in terms of their perihelion distance r_{min} and eccentricity e, by the equation

$$\blacktriangleright \qquad r_N(\phi) = \frac{r_{min}(1+e)}{1+e\cos\phi}. \qquad (5.3.7)$$

If e is less than unity, the orbits are bound and elliptical in shape (figure 41). By substituting into (5.3.6) (with $c \rightarrow \infty$), it is easily verified that r_{min} and e are related to $\mathscr{E}$ and B by

$$\blacktriangleright \qquad \left.\begin{array}{l} B^2 = GMr_{min}(1+e), \\[6pt] \mathscr{E} = -GM(1-e)/2r_{min}. \end{array}\right\} \qquad (5.3.8)$$

We are interested in the correction to the elliptical orbits (5.3.7) caused by the relativistic term $2GM/c^2r(\phi)$ in (5.3.6). For Mercury, the value of this term is only about 10^{-7}, so that it is necessary to calculate only the 'first-order corrections' to the Newtonian orbit $r_N(\phi)$. We write the relativistic orbit as

$$r(\phi) = r_{min}(1+e)/(1+e\cos\phi+\alpha(\phi)),$$

where $\alpha(\phi)$ is a small function which we require only in the first approximation. (We write $\alpha(\phi)$ in the denominator as a correction to $e\cos\phi$ because it is the term $\cos\phi$ in $r_N(\phi)$ that makes the Newtonian orbit a closed curve: when $\phi = 2n\pi$, $r_N(\phi)$ is always equal to the perihelic distance r_{min}.) To find $\alpha(\phi)$ we substitute into the exact orbit equation (5.3.6), neglecting terms of order $\alpha^2(\phi)$, $(d\alpha/d\phi)^2$ etc. This gives

$$\frac{B^2}{2\rho^2}\left[-2\frac{d\alpha}{d\phi}e\sin\phi+e^2\sin^2\phi+1+2e\cos\phi+e^2\cos^2\phi \right.$$

$$\left. +2\alpha(\phi)(1+e\cos\phi)-\frac{2GM}{c^2\rho}(1+e\cos\phi)^3 \right]$$

$$-\frac{GM}{\rho}(1+e\cos\phi+\alpha(\phi)) = \mathscr{E},$$

where, for convenience, we have written

$$\rho \equiv r_{\text{min}}(1+e).$$

The 'zeroth-order terms' are those not involving $\alpha(\phi)$ or $1/c^2$; setting them equal to zero simply gives the already-found Newtonian formulae (5.3.8) for B^2 and $\mathscr{E}$. The vanishing of the sum of the remaining 'first-order terms' gives an equation for $\alpha(\phi)$, namely

$$-\frac{d\alpha(\phi)}{d\phi} e \sin \phi + \alpha(\phi) e \cos \phi = \frac{GM}{c^2\rho}(1+e \cos \phi)^3.$$

It can easily be verified that the solution is

$$\alpha(\phi) = \frac{GM}{c^2\rho}\left[(3+2e^2)+\frac{(1+3e^2)}{e}\cos \phi - e^2 \cos^2 \phi + 3e\phi \sin \phi\right].$$

The first two terms are insignificant contributions of the same form as those already existing in the denominator of $r_N(\phi)$, and their only effect is to change slightly the interpretation of r_{min} and e. The third term is also unimportant: it causes a slight periodic variation in the position of perihelion. The important term is the last one, because the occurrence of ϕ on its own (that is, not in a periodic function) causes a *cumulative* effect after many revolutions, and this is the observed precession. Thus the orbit is

$$r(\phi) = \rho/\{1+e[\cos \phi+(3GM/c^2\rho) \phi \sin \phi]\},$$

or

▶ $$r(\phi) = r_{\text{min}}(1+e)/\{1+e \cos [(1-3GM/c^2r_{\text{min}}(1+e)) \phi]\}, \quad (5.3.9)$$

where the trigonometric approximation is correct to first order in $GM/c^2\rho$, which is the accuracy to which we have been working. Perihelia occur when the cosine is unity, that is, when

$$\phi(1-3GM/c^2r_{\text{min}}(1+e)) = 2\pi n,$$

where n is any integer. This may be written as

$$\phi = 2\pi n+6\pi nGM/c^2r_{\text{min}}(1+e).$$

Therefore the perihelion steadily advances (that is, the orbit rotates as a whole in the same direction as the planet moves) and the angular precession $\Delta\phi$ per revolution is

$$\Delta\phi = 6\pi GM/c^2r_{\text{min}}(1+e).$$

The centennial precession $\Delta\phi^{100}$ is

▶ $$\Delta\phi^{100} = 6\pi GM\mathscr{N}/c^2r_{\text{min}}(1+e), \quad (5.3.10)$$

Table 3

| | $\Delta\phi^{100}$ (seconds of arc) | |
Test body	Observation	General relativity
Mercury ☿	43.11 ± 0.45	43.03
Venus ♀	8.4 ± 4.8	8.6
Earth ⊕	5.0 ± 1.2	3.8
Icarus	9.8 ± 0.8	10.3

where $\mathcal{N}$ is the number of revolutions per century. (Recall from section 3.2 that $\Delta\phi^{100}$ is the precession left unexplained by Newtonian mechanics when the much larger effects caused by the other planets have been taken into account.)

Only for the planets Mercury, Venus and the Earth, and the asteroid Icarus, is r_{min} small enough, and $\mathcal{N}$ large enough, for $\Delta\phi^{100}$ to be measurable. The results are as shown in table 3. The large uncertainty in the measured precession of Venus arises from the near-circularity of the orbit (e is only 0.0068), which makes it difficult to locate the perihelion. This group of results forms a most impressive verification of the predictions of general relativity.

As we mentioned in section 3.4, it is possible to construct a 'semi-relativistic' theory of gravitation; this is done simply by adding the gravitational potential $-GM/r$ to the 'energy per unit mass' as given by special relativity. Thus if $\mathbf{p}$ is the momentum of the test particle, defined by $m\,dr/d\tau$ (m is the mass of the particle), then we have, as the energy equation,

▶ $$\sqrt{(p^2c^2+m^2c^4)}-GMm/r = mc^2+m\mathscr{E}, \qquad (5.3.11)$$

where $m\mathscr{E}$ is the excess energy of the particle over its rest energy mc^2. Using the relation

$$p^2 = m(r'^2+r^2\phi'^2),$$

and the conservation of angular momentum (5.3.4), together with (5.3.11), we obtain

$$\frac{B^2}{2r^2}\left(1+\frac{dr/d\phi^2}{r^2}\right) - \frac{GM}{r} = \mathscr{E}+\frac{(\mathscr{E}+GM/r)^2}{2c^2}.$$

This should be compared with the equation (5.3.6) from general relativity. The form of the term giving the correction to Newtonian mechanics

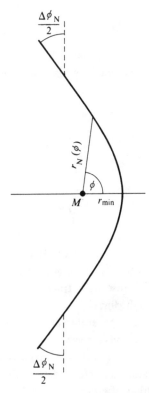

Figure 42. Notation for Newtonian light track.

is completely different and the resulting precession is only *one-sixth* of that given by equation (5.3.10); thus this flat-space theory is quite unable to explain the data of table 3.

5.4 Deflection of light

In section 3.6 we saw that Einstein's principle of equivalence implies that light is deflected in a gravitational field. As was first realised by Soldner in 1801, a deflection is also predicted by Newtonian mechanics, and we begin by calculating its value. The formula (5.3.7) for the Newtonian orbit corresponds to unbound, hyperbolic paths if the eccentricity e exceeds unity; then, according to (5.3.8), the 'energy' $\mathscr{E}$ is positive. The asymptotes, where $r = \infty$ (figure 42), correspond to angles

$$\phi = \pm(\pi/2 + \Delta\phi_N/2),$$

where $\Delta\phi_N$ is the total Newtonian deflection of the ray, given by

$$\cos\phi = -1/e, \quad \text{i.e. } \sin\Delta\phi_N/2 = 1/e.$$

For light, the known speed c at infinity gives

$$\mathscr{E} = c^2/2$$

since this is the known value of the 'energy per unit mass' far from the gravitating centre. This enables us to find the eccentricity via the second equation (5.3.8); it is

$$e = 1 + 2r_{min}\mathscr{E}/GM = 1 + c^2 r_{min}/GM \approx c^2 r_{min}/GM,$$

since $c^2 r_{min}/GM \gg 1$ in all practical cases. Thus e is very large, and $\Delta\phi_N$ is very small, so that we have,

$$\sin(\Delta\phi_N/2) \approx \Delta\phi_N/2 = 1/e,$$

that is

▶ $$\Delta\phi_N = 2GM/c^2 r_{min}. \qquad (5.4.1)$$

For light grazing the surface of the Sun, $r_{min} = R_\odot$ and $M = M_\odot$, giving $\Delta\phi_N = 0.875''$. If we had used special relativity instead of Newtonian mechanics, we would have found $\Delta\phi = 0$, because light tracks are null geodesics, and in flat spacetime these are straight lines.

Now we calculate the deflection $\Delta\phi$ according to general relativity. Light travels along null geodesics, on which all separations $\Delta\tau$ vanish. This does not mean we have to do the 'extremalising' of the last section all over again, because null geodesics are special cases, characterised by infinite values of the 'constants of the motion' A and B of equations (5.3.3) and (5.3.4). However, A/B is a finite constant, because $d\phi/dt$ is finite. Thus the general orbit equation (5.3.6) reduces for null geodesics to

▶ $$[dr(\phi)/d\phi]^2/r^2(\phi) + 1 - 2GM/c^2 r(\phi) = Dr^2(\phi), \qquad (5.4.2)$$

where D is a new constant, given by $A^2 c^2/B^2$.

If there were no attracting mass, M would be zero and the path would be a straight line $r_s(\phi)$ (figure 43), which we write in the form

$$r_s(\phi) = r_{min}/\cos\phi.$$

substitution into (5.4.2) (with $M = 0$) gives $D = 1/r_{min}^2$. When M is not zero, we write a 'trial solution' as

$$r(\phi) = r_{min}/(\cos\phi + \beta(\phi)),$$

and calculate the correction term $\beta(\phi)$ in lowest order. Substituting in (5.4.2), neglecting terms of order $\beta^2(\phi)$, and using the already-found value for D, we get, after a little reduction, the following equation for $\beta(\phi)$:

$$\beta(\phi)\cos\phi - (d\beta(\phi)/d\phi)\sin\phi = (GM/c^2 r_{min})\cos^3\phi.$$

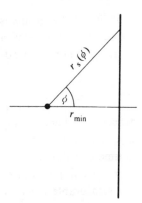

Figure 43. Straight light track when there is no central mass.

The solution is easily verified to be

$$\beta(\phi) = (2GM/c^2r_{min}) - (GM/c^2r_{min}) \cos^2 \phi.$$

The asymptotes lie near $\phi = \pm \pi/2$, so that the second term is negligible and we have, for the orbit,

▶ $$r(\phi) = r_{min}/(\cos \phi + 2GM/c^2r_{min}).$$ (5.4.3)

The directions of the asymptotes are given by

$$-\cos \phi = \sin (\Delta\phi/2) \approx \Delta\phi/2 = 2GM/c^2r_{min},$$

so that the deflection $\Delta\phi$ is

▶ $$\Delta\phi = 4GM/c^2r_{min}.$$ (5.4.4)

This is exactly twice the Newtonian value (5.4.1). For light grazing the Sun, $\Delta\phi = 1.75''$.

It is possible to measure the deflection of the light from a star only during a total eclipse of the Sun, because at other times the glare from the Sun makes observation difficult. The apparent position of the chosen star (relative to other stars) during an eclipse is compared with the position six months later when the star is in the night sky; the angular difference in apparent positions is the required deflection. In this way $\Delta\phi$ has been measured for about 400 stars. It is also possible to measure radio deflections, due to the fortunate circumstance of the annual occultation by the Sun of the QSO 3C 279, which happens to be a radio emitter of great power. None of the results of these experiments are compatible with the Newtonian prediction of 0.875″, because they all lie between 1.57″ and 2.37″. The average value of all the observed deflections is 1.89″, which agrees well with Einstein's prediction of 1.75″. (For 3C279 the result is 1.73″ ± 0.05″.) The first verification, during the eclipse of 1919, created a popular sensation, and one American newspaper carried the famous head-line 'Light caught bending'.

5.5 Radar echoes from planets

We have seen how measurements on the deflection of light passing near the Sun can be used to test the predictions of general relativity con-cerning the shape of the orbits of light rays. The time coordinate along the orbit was carefully eliminated to get an equation for r in terms of ϕ. However, modern radar techniques render directly measurable the time taken by a signal to travel along a path to a reflector, and back again. To detect relativistic effects the path is chosen to pass near the Sun, and a convenient reflector is a planet (or space probe) near 'superior conjunc-tion'; this means that the Sun lies between the Earth and the planet, almost on a straight line joining them. Rather surprisingly, general rela-tivity predicts that the echo would be *delayed* relative to an imaginary signal which travels out and back in a straight line with speed c: the radiation slows down as it passes the Sun. To see this, we use the Schwarz-schild metric (5.1.2) directly to calculate the coordinate speed of light travelling in the (r, ϕ) plane defined by $\theta = \pi/2$. For light, $\Delta\tau$ is zero. Thus, if the motion is purely *radial* ($\Delta\theta = \Delta\phi = 0$), the speed is

$$\Delta r/\Delta t = c(1 - 2GM/c^2r),$$

while for purely *transverse* motion ($\Delta r = \Delta\theta = 0$), the speed is

$$r\Delta\phi/\Delta t = c\sqrt{(1 - 2GM/c^2r)};$$

both speeds are less than c.

The coordinate time T between the emission of the signal and the re-ception of the echo is

$$T = 2[t(r_\oplus, r_{min}) + t(r_p, r_{min})],$$

where $t(r, r_{min})$ is the coordinate time taken for light to travel from r in to the radius of closest approach (figure 44), and $r_\oplus$ and r_p are the coordi-nate distance of the Earth and planet from the Sun. T is the coordinate time between two events on the Earth, that is, at essentially fixed r, θ, ϕ. The corresponding proper time, which is what our clocks would measure, is smaller by a factor

$$\sqrt{(1 - 2GM_\odot/c^2r_\oplus)};$$

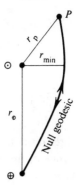

Figure 44. Notation for radar time delay theory.

this differs from unity by only 10^{-8}, which corresponds to a few microseconds in experiments involving Mercury and Venus. The relativistic time-delay excess amounts to about 100 microseconds, so we shall ignore the time dilation factor in what follows. If light travelled in straight lines with speed c, we would have $t(r, r_{\min}) = t^0(r, r_{\min})$ where

$$t^0(r, r_{\min}) = \sqrt{(r^2 - r_{\min}^2)}/c.$$

To calculate $t(r, r_{\min})$ relativistically, we write the Schwarzschild metric (5.1.2) for a null separation in the form

$$\blacktriangleright \qquad 0 = 1 - \frac{2GM}{c^2 r} - \frac{1}{c^2}\left[\frac{(dr/dt)^2}{1 - 2GM/c^2 r} + r^2\left(\frac{d\phi}{dt}\right)^2\right]. \qquad (5.5.1)$$

To eliminate $d\phi/dt$ we use the geodesic equations of motion (5.3.3) and (5.3.4), whose quotient gives

$$\frac{r^2\phi'}{(1 - 2GM/c^2 r)t'} = \frac{r^2(d\phi/dt)}{1 - 2GM/c^2 r} = \frac{B}{A} \equiv D.$$

Thus (5.5.1) becomes

$$\blacktriangleright \qquad 0 = 1 - \frac{2GM}{c^2 r} - \frac{1}{c^2}\left[\frac{(dr/dt)^2}{1 - 2GM/c^2 r} + \frac{D^2}{r^2}\left(1 - \frac{2GM}{c^2 r}\right)^2\right]. \qquad (5.5.2)$$

We can relate D to $r_{\min}$ by realising that dr/dt vanishes when $r = r_{\min}$, since the direction of the light ray at that point has no radial component. This leads to

$$D^2 = c^2 r_{\min}^2/(1 - 2GM/c^2 r_{\min}),$$

so that (5.5.2) gives

$$\frac{dr}{dt} = c\left(1 - \frac{2GM}{c^2 r}\right)\sqrt{\left[1 - \frac{r_{\min}^2(1 - 2GM/c^2 r)}{r^2(1 - 2GM/c^2 r_{\min})}\right]}.$$

By integration, we easily obtain

$$t(r, r_{\min}) = \frac{1}{c}$$

$$\times \int_{r_{\min}}^{r} \frac{dr}{(1 - 2GM/c^2 r)\,\sqrt{[1 - r_{\min}^2(1 - 2GM/c^2 r)/r^2(1 - 2GM/c^2 r_{\min})]}}. \tag{5.5.3}$$

As with the other tests of general relativity, this exact result is impenetrable, but it suffices to work out the integral to first order in the small quantity $GM/c^2 r$. The necessary expansion of the integrand is not quite straightforward, but leads to the result

$$t(r, r_{\min}) \approx \frac{1}{c} \int_{r_{\min}}^{r} \frac{r\,dr}{\sqrt{(r^2 - r_{\min}^2)}} \left[1 + \frac{2GM}{c^2 r} + \frac{GM r_{\min}}{c^2 r(r + r_{\min})} \right].$$

The integrals are elementary, and give

$$t(r, r_{\min}) = \frac{1}{c} \left[\sqrt{(r^2 - r_{\min}^2)} + \frac{2GM}{c^2} \ln \left(\frac{r + \sqrt{(r^2 - r_{\min}^2)}}{r_{\min}} \right) \right.$$

$$\left. + \frac{GM}{c^2} \sqrt{\frac{r - r_{\min}}{r + r_{\min}}} \right]. \tag{5.5.4}$$

It is seen how the terms in GM/c^2 provide small corrections to $t^0(r, r_{\min})$. We require the *excess* time delay ΔT for the whole trip; this is defined as

$$\Delta T \equiv T - (2/c)\,\sqrt{(r_{\oplus}^2 - r_{\min}^2)} - (2/c)\,\sqrt{(r_{\mathrm p}^2 - r_{\min}^2)}.$$

In practical cases $r_{\min}$ is approximately the Sun's radius, so that $r_{\oplus} \gg r_{\min}$ and $r_{\mathrm p} \gg r_{\min}$; therefore the square roots in the correction terms in (5.5.4) can be simplified, and the final result is

$$\Delta T \approx (4GM_{\odot}/c^3)\,(\ln(4r_{\oplus} r_{\mathrm p}/r_{\min}^2) + 1). \tag{5.5.5}$$

For echoes from Mercury which just graze the Sun, this formula gives $\Delta T = 2.4 \times 10^{-4}$ s, which is easily measurable.

Several experiments have been performed, principally by Shapiro and his co-workers; in every case the predictions of general relativity are confirmed. We show as an example the results (figure 45) for the superior conjunction of Venus in 1970. Two points deserve emphasis: first, we have here agreement not just with a *single value* of a physical quantity but with a *functional form* as well; second, the results are totally incompatible with the Newtonian excess time delay $\Delta T_{\mathrm N}$, derived on the assumption that light is a stream of material particles whose speed at $r = \infty$ is c. The formula for $\Delta T_{\mathrm N}$ is

$$\Delta T_{\mathrm N} \approx -(2GM_{\odot}/c^3)\,(\ln(4r_{\oplus} r_{\mathrm p}/r_{\min}^2) - 2). \tag{5.5.6}$$

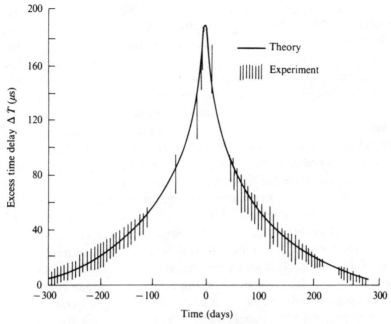

Figure 45. Radar time delay of signals reflected from Venus and passing close to the Sun: comparison of theory (equation (5.5.5)) with experiment. (By kind permission of Professor I. I. Shapiro.)

This predicts that ΔT_N is negative in all practical cases, as we would expect from Newtonian mechanics, according to which light should travel faster near the Sun.

5.6 Black holes

In the last four sections, the effects of general relativity have been very small corrections to the prediction of Newtonian mechanics. Now we discuss a situation with no exact Newtonian analogue, where general relativity dominates. We begin by noting that in the Schwarzschild separation formula (5.1.2) the metric coefficient g_{11} (i.e. g_{rr}) becomes infinite if r equals the 'Schwarzschild radius' r_S, defined as

▶ $$r_S \equiv 2GM/c^2. \tag{5.6.1}$$

In most cases r_S is far smaller than the actual radius of the body concerned: for the Sun, r_S is 2.95 km, while for the Earth r_S is 8.86 mm. Now, the Schwarszchild metric applies only in the space outside the gravitating body; inside a different formula holds, and g_{11} is not singular at r_S. It might seem, therefore, that the 'Schwarzschild singularity' is as unphysical as the apparent singularity at the origin in the Newtonian gravitational potential $-GM/r$, a formula which also applies outside the body only. However, astrophysicists have calculated that the ultimate stage in the evolution of stars several (~ 3) times more massive than the Sun would be 'gravitational collapse'. In this process, the star has exhausted its nuclear fuel and no longer radiates, and the pressure of its matter cannot resist the gravitational self-attraction of the star, which therefore shrinks down within its Schwarzschild sphere at $r = r_S$. Such collapsed stars, and other objects smaller than their Schwarzschild radii, are called *black holes*.

Black holes are very strange objects. Their most astonishing property is that anything inside the Schwarzschild sphere must fall into the centre $r = 0$. This applies to the matter constituting the star whose collapse formed the black hole, and implies that the collapse continues until the star is a point singularity at $r = 0$. (The implication holds if general relativity remains valid in the unimaginably dense hot conditions near the end point of the collapse. If quantum or other effects intervene, the singularity might be avoided. However, no consistent theory of 'quantum gravity' has yet been constructed, and this is still very much an open question.) The 'collapse' property also applies to any light emitted by the star when its radius is less than r_S, and implies that no light can escape, so that these objects cannot be seen from outside – hence the name 'black holes'. (We are not saying that black holes exert no influence on their surroundings; their gravitational effects on external bodies are perfectly normal, because these effects arise from the Schwarzschild metric for $r > r_S$.)

Now we must prove the truth of these startling assertions. First we recall that two neighbouring events cannot lie on the world line of a particle or light ray if $\Delta\tau^2$ is negative (cf. section 4.1). For example, in the flat space-time whose separation formula is given by (4.1.3), we cannot have a given particle existing in two places at the same time because this would imply $\Delta t = 0$, $|\Delta\mathbf{r}| \neq 0$, so that $\Delta\tau^2$ would be negative. In the Schwarzschild spacetime (5.1.2) this implies that no particle can be stationary inside a black hole. The argument runs as follows: for a particle at rest,

$$\Delta r = \Delta\theta = \Delta\phi = 0;$$

but when $r < r_S$ the coefficient $g_{00} = 1 - 2GM/c^2 r$ of Δt^2 is negative, so that $\Delta\tau^2$ is negative for such events. Outside the black hole, g_{00} is positive, so that it *is* possible for a body to be at rest (of course such a body does not pursue a geodesic world line; other forces must act to keep it stationary – a good example is ourselves, at rest on the Earth with our geodesic free fall prevented by upward forces from the ground).

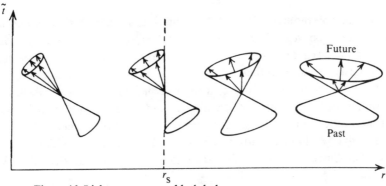

Figure 46. Light cones near a black hole.

Thus there is no 'statics' inside a black hole, only dynamics. Do particles move inwards or outwards? To answer this we examine the continuity of the 'null cones'. These are the double (hyper-)cones joining an event P to those neighbouring events corresponding to zero separation. The timelike geodesics from P (possible world lines), for which $\Delta\tau^2 > 0$, all lie within the cones. The lines in one cone point into P's future, while those in the other point into P's past. Figure 46 shows cones in the plane of r and a new time coordinate (not singular at r_S) given by

$$\tilde{t} = t + (2GM/c^3)\ln|rc^2/2GM - 1|.$$

The Schwarzschild formula now gives the equation of the cones as

i.e.
$$0 = (1 - 2GM/c^2r)\,\Delta\tilde{t}^2 - 4GM\Delta r\Delta\tilde{t}/c^3r - \Delta r^2\,(1 + 2GM/c^2r)/c^2$$

▶ $$\frac{d\tilde{t}}{dr} = -\frac{1}{c} \text{ and } \frac{1}{c}\frac{1 + 2GM/c^2r}{1 - 2GM/c^2r}. \qquad (5.6.2)$$

For $r \gg r_S$ the future null cones point upwards, that is, the world lines correspond to increasing coordinate time $\tilde{t}$. Some lines in the future cones point inwards, and some outwards, that is, particles can travel either towards or away from the central mass. As we approach the black hole, however, the future cones lean inwards, until when $r < r_S$ all world lines point inwards towards $r = 0$. Thus the collapse of all the matter and radiation inside a black hole is inevitable. Any attempt to try to prevent it by applying forces (for instance the compressive stresses generated as a star collapses) is as futile as an attempt by us to try to travel in our own past by applying forces – it is impossible to turn back the hands of time.

The fact that g_{rr} becomes infinite at the Schwarzschild radius does not mean that spacetime itself is singular there. The curvature (4.4.1) behaves in a perfectly smooth manner at $r = r_S$. In fact r_S is merely a *coordinate singularity*, and other coordinate systems have been found that describe Schwarzschild spacetime in such a way that all g_{ij} vary smoothly through the black-hole radius (cf. (5.6.2)). The situation is similar to that in ordinary flat space when we use polar coordinates; then $g_{\theta\theta}$ and $g_{\phi\phi}$ are infinite when r is infinite (equation (4.1.2)). The centre $r = 0$ is however a genuine singularity of the spacetime itself, irrespective of any coordinates.

Now we study the history of a body falling radially into a black hole from rest at infinity. First we calculate its radial coordinate r as a function of the proper time τ, that is, the time as measured by a clock falling with the body. We have already worked out the necessary equations in section 5.3. Since there is no angular motion, ϕ' and θ' vanish and (5.3.5) becomes

$$\tfrac{1}{2}[r'(\tau)]^2 - GM/r(\tau) = (A^2 - 1)\,c^2/2 \equiv \mathscr{E},$$

– just the Newtonian equation! Since $r'(\infty) = 0$, we must have $\mathscr{E} = 0$, $A^2 = 1$, so that

$$r(\tau)(r'(\tau))^2 = 2GM.$$

If the travelling clock is set so as to read $\tau = 0$ at the moment of collapse at $r = 0$, then the solution of this equation is

▶ $$r(\tau) = (-3\tau/2)^{\frac{2}{3}}\,(2GM)^{\frac{1}{3}}. \tag{5.6.3}$$

Thus the body falls in smoothly from $\tau = -\infty$. Nothing special happens when it crosses r_{S} at the time

$$\tau_{(r = r_{\mathrm{S}})} = -2r_{\mathrm{S}}/3c.$$

The collapse is very rapid: for a black hole of solar mass the body lives for only 10^{-5} s after crossing r_{S} before being destroyed in the central singularity.

No observer outside the black hole can see the final stages of collapse because, as we have shown, no signal, optical or material, can escape from $r < r_{\mathrm{S}}$ to $r > r_{\mathrm{S}}$. What then *does* an external observer see as a star collapses, or as a body emitting light falls into an already existing black hole? First we must draw a proper spacetime diagram of the collapse. This means that in the trajectory $r(\tau)$ (equation (5.6.3)) we must replace τ by t using (5.3.3) with $A = 1$ ($t = \tau$ at $r = \infty$ in this case since the body falls from rest). Thus we have

$$dt = \frac{d\tau}{1 - r_{\mathrm{S}}/r(\tau)} = \frac{r\,dr(d\tau/dr)}{r - r_{\mathrm{S}}}$$

$$= -\frac{r^{\frac{3}{2}}\,dr}{\sqrt{(2GM)}\,(r - r_{\mathrm{S}})},$$

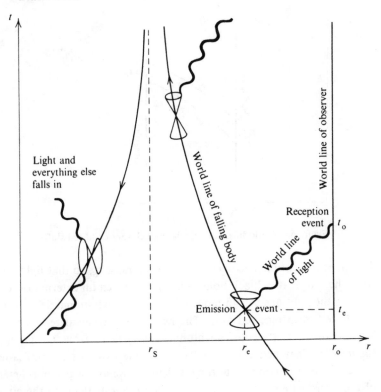

Figure 47. World lines near a black hole

an equation whose solution is

$$t = \frac{r_S}{c}\left[-\frac{2}{3}\left(\frac{r}{r_S}\right)^{\frac{3}{2}} - 2\left(\frac{r}{r_S}\right)^{\frac{1}{2}} + \ln\left|\frac{1+(r/r_S)^{\frac{1}{2}}}{1-(r/r_S)^{\frac{1}{2}}}\right| \right], \qquad (5.6.3a)$$

corresponding to the world line sketched on figure 47. We see that as the body approaches r_S, t becomes infinite, so that with respect to the coordinate clocks, sycnhronised with those at $r = \infty$, the infalling body appears 'frozen' at r_S. (The peculiar form of the world line for $r < r_S$ does not mean that the body travels backwards in time; it is an artefact of this particular coordinate system.)

Next we follow the history of the light emitted at various stages of the collapse, since it is by means of its light that we obtain information about the body. Light emitted at the event t_e, r_e travels radially outwards along a null geodesic to reach a distant observer at the event t_o, r_o (figure 47). Along such a world line r and t are related by (5.5.1), which gives

$$t_o - t_e = \frac{1}{c}\int_{r_e}^{r_o} \frac{dr}{1 - 2GM/c^2 r} = \frac{r_o - r_e}{c} + \frac{2GM}{c^3}\ln\left(\frac{r_o - r_S}{r_e - r_S}\right). \qquad (5.6.4)$$

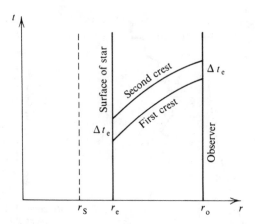

Figure 48. World lines for calculating Schwarzschild red shift.

This light travel time becomes infinite when $r_e = r_S$, so that light emitted from the 'edge' of a black hole would never reach an external observer.

Since the light has to climb out of a strong gravitational field, it will be red-shifted when received at r_0. The next step is to calculate this red shift. Two wave crests (ticks of an atomic clock) leave r_e at t_e and $t_e + \Delta t_e$ (figure 48). They arrive at t_0 and $t_0 + \Delta t_e$; the two Δt's are the same because of the way coordinate time is defined (we are neglecting the effects of the coordinate speed of the body when $r \approx r_S$). Between the emission events, the proper time $\Delta \tau_e$ is, from (5.1.2),

$$\Delta \tau_e \equiv 1/\nu_e \equiv \lambda_e/c = \Delta t_e \sqrt{(1 - 2GM/c^2 r_e)},$$

where ν_e and λ_e are the frequency and wavelength of the source as measured in its rest frame. For the observer at r_0 precisely analogous relations hold:

$$\Delta \tau_0 \equiv 1/\nu_0 \equiv \lambda_0/c = \Delta t_e \sqrt{(1 - 2GM/c^2 r_0)}.$$

Thus the red shift z (equation (2.3.1) is simply

$$\blacktriangleright \quad z = \sqrt{\frac{1 - 2GM/c^2 r_0}{1 - 2GM/c^2 r_e}} - 1 = \sqrt{\frac{r_e(r_0 - r_S)}{r_0(r_e - r_S)}} - 1. \tag{5.6.5}$$

As $r_e \to r_S$, z becomes infinite, so that the light emitted by a star just as it collapses into a black hole is infinitely reddened and hence unobservable. If $r_e \gg r_S$ and $r_0 \gg r_S$, this exact red shift reduces to our earlier approximate results (2.3.4) (derived using energy conservation for photons, and valid when $r_0 \to \infty$), and (3.6.3) (derived using the equivalence principle and valid when $r_0 \approx r_e$). Infinite red shift means infinite time dilation, so that black holes are the fountain of youth – anyone sitting at rest on the 'edge' of a black hole would remain forever young in comparison with people far outside the hole, although he would of course age in his own rest frame.

We have shown that an external observer always will receive some light from a collapsed star, namely that emitted in the last moments before it contracted down through its Schwarzschild sphere. Therefore the formation of a black hole is not signalled by the *instantaneous* disappearance of the star's light. The light will get continuously dimmer, because the energy of each photon is reduced by the red shift factor $(1+z)$, and also because the photons will arrive separated by longer and longer intervals. These two effects imply that the apparent luminosity $l(t_0)$ of the star seen by the observer at coordinate time t_0 is less than the luminosity l_1 of a similar star at the same distance in a flat spacetime, by a factor

$$l(t_0)/l_1 = 1/(1+z)^2 = (r_e - r_S)\, r_0/(r_0 - r_S)\, r_e.$$

Now, t_0 is related to r_e by (5.6.4), which simplifies when $r_e \approx r_S$ to

$$t_0 \approx (r_S/c) \ln [(r_0 - r_S)/(r_e - r_S)],$$

so that the luminosity ratio tends to the form

$$l(t_0)/l_1 \overset{t_0 \to \infty}{\approx} (r_0/r_S) \exp (-ct_0/r_S).$$

Therefore the received power diminishes exponentially. This means that the apparent magnitude m increases linearly, according to the law (cf. equation (2.2.6))

$$m = 2.5 \log_{10} (l_1/l(t_0)) + \text{const.}$$
$$= 1.09 ct_0/r_S + \text{const.}$$

This is very rapid: the time taken for a collapsing star of solar mass to dim from first magnitude to the faintest magnitude detectable in our largest telescopes ($m = 22.7$) is only about 2×10^{-4} s. Thus for all practical purposes a collapsing star *does* appear to switch off like a light.

The black holes we have discussed so far are of a very special kind: they are non-rotating. However, most stars possess angular momentum, and their collapse will produce *rotating black holes*. To see that the rotation of a mass affects its external gravitational field, consider Mach's principle: this tells us that the inertial frames in a region of spacetime are determined by all the matter in the universe. The distant matter usually dominates, as we saw in section 3.5. But large masses nearby must have *some* effect, and if they are rotating they will tend to drag the inertial frames around with them, so that an inertial observer will see the distant stars rotate. The effect will usually be very small; to estimate it we assume that the magnitude ω

of the angular velocity of the dragged inertial frames (relative to the stars) is proportional to the magnitude J of the angular momentum of the nearby body, which is at a distance r. Only G, c, J, and r can be involved in the formula for ω, and we must have

$$\omega = kGJ/c^2r^3,$$

where k is a dimensionless constant. Of course the angular momentum and angular velocity are actually vectors, and it turns out that this leads to a dependence of k on the polar angle θ of the observer – in the 'equatorial' plane the vectors are antiparallel and $k = -1$. If we put in the figures for the Earth's surface, we find $\omega \sim 0.1''$ per year. Thus the axis of a gyroscope, or the plane of oscillation of a Foucault pendulum, will not quite be fixed relative to the distant stars, but will precess slowly with a period of about 10^7 years.

Near a black hole, however, this dragging of inertial frames can become very rapid. If the star inside the black hole had an angular velocity Ω at the instant it contracted through its Schwarzschild radius r_S, and was uniformly dense, then a simple calculation (using the moment of inertia for a homogeneous sphere) gives

$$\omega/\Omega = r_S^3/5r^3.$$

Thus observers who are not rotating (relative to the stars) and who are situated just outside the hole will feel dizzy as the inertial frames whirl around them. For collapse of an object of solar mass and angular momentum, our approximate formula gives $\omega \sim 10^5$ rad s^{-1} for $r \approx r_S$.

The rigorous treatment of rotating black holes is based on general relativity. Kerr has generalised the Schwarzschild separation formula to include the effects of the angular momentum $\mathbf{J}$ of the collapsed object. The analysis is fascinating, but the details are complicated, so we shall simply describe the results. Let the direction of $\mathbf{J}$ define the positive axis of polar coordinates, so that the original star's angular velocity corresponds to increasing azimuth angle ϕ. If the magnitude $J > GM^2/c$ no black hole forms; roughly speaking, we can say that centrifugal forces prevent collapse. If $J < GM^2/c$ a black hole does form, with Schwarzschild radius given not by (5.6.1) but by

$$\blacktriangleright \qquad r_S = GM/c^2 + \sqrt{[(GM/c^2)^2 - (J/Mc)^2]}. \qquad (5.6.6)$$

This has the same property as a non-rotating black hole, namely that matter and radiation inside r_S must be falling in – the future null cones point inwards and there is no escape.

But this is not all: there is another surface, non-spherical this time,

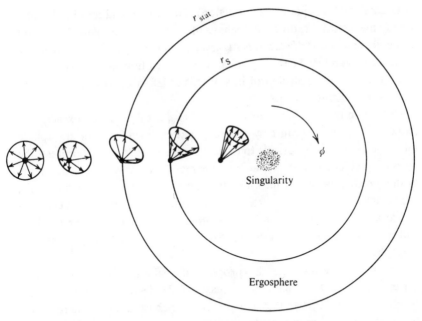

Figure 49. Future light cones near rotating black hole.

outside the Schwarzschild sphere, called the *static limit*, whose radius r_{stat} is given by

$$\blacktriangleright \qquad r_{stat} = GM/c^2 + \sqrt{[(GM/c^2)^2 - (J/Mc)^2 \cos^2 \theta]}. \qquad (5.6.7)$$

Inside this surface, it is impossible to remain at rest. *Angular motion is compulsory*; the future null cones lean over (figure 49) towards $+\phi$ as r decreases, until when $r < r_{stat}$ no 'future' world line points towards decreasing ϕ. Within the static limit the dragging of inertial frames is so violent that it is impossible to remain non-rotating relative to distant stars. By contrast, *radial* motion is not compulsory; it is possible to cross the static limit from above or below, or to rotate within r_{stat} with constant r. Only within r_S is radial motion restricted to infall (figure 49).

The central singularity of a rotating black hole is very complicated, and the final stages of collapse are not well understood. It seems conceivable that some kind of re-explosion might occur but this is most peculiar, because the expanding object could reach some given r at the same time as the contracting object! In other words matter could have closed *timelike world lines* – a gross violation of causality. One bizarre possible way out of this dilemma is suggested by the mathematics: the expansion occurs into 'another' universe (related to our own like the positive and negative 'branches' of the square root function). Objects exploding out through their Schwarzschild radii would behave like time-reversed black holes, and hence are called *white holes*. It is suggested that some supernovae might be white holes, produced either by collapse of a black hole in another universe or by the delayed explosion of bits of matter left over from the big bang at the birth of this universe.

The region of spacetime between r_S and r_{stat} is called the *ergosphere*, the name being chosen because by dropping matter down from the ergosphere into a black hole it is possible to *extract energy from it*. In this process, the mass of the black hole decreases, and the recoil increases the kinetic energy of the receding vehicle from which the matter is dropped. To achieve this, the vehicle must eject the matter into an orbit with negative total energy – i.e. into a very strongly bound orbit – otherwise the mass of the black hole will increase rather than decrease. Such orbits can exist only within the ergosphere.

It might be thought that this process will make the black hole smaller since, according to (5.6.6), r_S decreases if M decreases; eventually we would have $J > GM^2/c$ and the protective Schwarzschild sphere of the black hole would be destroyed, exposing the fiery singularity within. However, this could not occur, because detailed analysis of possible orbits of particles near r_S, as well as a more general argument, shows that whatever happens to a black hole *its proper surface area A cannot decrease*. From the Kerr separation formula, A can be calculated to be

$$\blacktriangleright \quad A = 4\pi(r_S^2 + (J/Mc)^2)$$
$$= (8\pi GM/c^2)\{GM/c^2 + \sqrt{[(GM/c^2)^2 - (J/Mc)^2]}\}. \quad (5.6.8)$$

Reducing M does tend to reduce A and make the black hole smaller, but reducing J tends to increase A. Thus for the black hole power station to deliver energy the matter must be injected in such a way as to reduce the angular momentum of the black hole. In brief, the energy comes from the rotational energy of the black hole.

Three very powerful theorems about black holes have been proved under wide conditions (although at the time of writing the proofs are not completely general). The first is that gravitational collapse always ends in some kind of singularity, even if the collapsing body is not spherically symmetric. The second theorem is that the rest of the universe is always shielded from the singularities by the development of a Schwarzschild sphere which prevents radiation and matter escaping (this sphere is one kind of 'event horizon' – see section 6.2); the non-decreasing area property discussed in the last paragraph is one of the mechanisms preventing the formation of 'naked singularities.' The third theorem is that once a black hole has been formed the only properties of the original body that survive to influence the external spacetime are the mass M, the angular momentum $\mathbf{J}$, and the electric charge Q. This spacetime requires a slight generalisation of the Kerr metric, to include the effect of Q; it is called the Kerr–Newman metric. All other properties of the body – shape, mechanical strength, electric dipole moment, magnetic multipole moments etc. – are destroyed in the collapse. Thus black holes are completely specified by M, $\mathbf{J}$, and Q; this has led to the theorem being expressed in the following way: 'A black hole has no hair.' These theorems must be used with care; the assumption of causality is built into their proofs and it is hard to see exactly how they apply even to the relatively simple case of rotating collapsing objects.

Finally we ask the important question: Do black holes actually exist? The possibility that they might was considered long before the days of relativity. In 1798 Laplace wrote: 'A luminous star, of the same density as the Earth, and whose diameter would be two hundred and fifty times larger than the Sun, would not, in consequence of its attraction, allow any of its rays to arrive at us; it is therefore possible that the largest luminous bodies in the universe may, through this cause, be invisible.'

At the present time the best hope seems to be to search for black holes belonging to a binary star system whose other component is visible. The visible star can be identified as part of a binary because periodic Doppler shifts in its light reveal its orbital motion about the centre of mass. The mass M of the invisible component can be inferred as follows. The spectral type of the visible star enables its mass M_1 to be estimated, while its orbital period and orbital speed enable the 'mass function' $M^3/(M+M_1)^2$ to be determined, and thence M. If M turns out to exceed about $3\,M_\odot$ (the 'critical mass' for collapse), the invisible object might be a black hole. Without further information we cannot say that it *is* a black hole, because, for example, it might be a very small (i.e. $M \ll 3M_\odot$) cold star surrounded by a massive halo. However, if the two stars are very close together another effect occurs which could help to decide the issue. This is the siphoning off of matter from the visible star by the intense gravitational field of the invisible star. As this matter falls it gets compressed and heated, and it has been calculated that this process should generate X-rays of a characteristic sort. The observation of such X-rays does not by itself guarantee a black hole, because similar radiation is expected from *neutron stars* which are stars with $M \lesssim 3M_\odot$ whose gravitational contraction can be stopped (by pressure) just short of collapse (i.e. $r > r_{\mathrm{s}}$). It is necessary also to know the mass, and for this we need a visible companion, as we have explained.

So far one strong contender for the title of First Black Hole has emerged. This is *Cygnus* X1, which is a powerful emitter of X-rays. The strength of these X-rays fluctuates violently on a scale of 0.1 s, which suggests that the source is very compact (diameter < 0.1 light-second $\approx 3 \times 10^7$ m). The companion star which is steadily being swallowed up has been identified optically; its observed orbital motion leads to the inference that the X-ray source has a mass of about 14 $M_\odot$. Any radiating star with this mass would have an absolute luminosity far in excess of the Sun's, and would be clearly visible at the distance of this binary system (~ 2 kpc). Any star with this mass which is no longer radiating must have undergone complete gravitational collapse. Therefore Cygnus X1 is a black hole. (Some astronomers dispute this interpretation, and are searching for alternatives.)

Astrophysical arguments suggest that as many as one per cent of stars might by now have evolved into black holes (that is, 10^9 in our galaxy alone). In addition, giant black holes, with masses hundreds or thousands of times greater than $M_\odot$, might exist at the centres of globular clusters and at the centres of galaxies; as these gobble up their surroundings intense bursts of gravitational radiation should be produced (see section 3.5), but these are still outside the limit of detection using current technology.

6 *Cosmic kinematics*

6.1 Spacetime for the smoothed-out universe

If general relativity is correct, spacetime is puckered and warped in an incredibly complicated way, on all scales from atoms up to at least clusters of galaxies. In cosmology it is customary as a first approximation to regard this rich detail as 'fine structure', which may be ignored when studying processes involving the universe as a whole. Thus we imagine the inhomogeneities of spacetime as being *averaged out* over distances of the order of 10^7 pc, and consider any curvature remaining as a large-scale property of the universe. If we could break the galaxies apart and distribute their matter uniformly throughout space, we would have precisely the kind of 'uniform model universe' dealt with by cosmologists. This procedure has honorable precedents elsewhere in physics; for example, in studying the non-turbulent flow of fluids it is rarely necessary to consider the detailed motion of all the atoms, since only their *average* density and velocity determine bulk behaviour.

However, it may be that the universe is inhomogeneous on all scales, like the velocity distribution in a turbulent fluid: we might have galaxies, clusters of galaxies, clusters of clusters of galaxies, and so on. This is the *hierarchical* model, proposed by Charlier in 1908. If the universe is in fact organised in this way, then no overall averaging is possible, and all uniform models are useless. Optical astronomers have searched for inhomogeneity in the distribution of galaxies, and their results indicate some superclustering on scales of 30–50 Mpc. However, it is possible with radio astronomy to reach out to much greater distances, and on these larger scales it does appear that the universe is at last homogeneous. Thus there is no compelling reason for us to consider the hierarchical model any further. Indeed, there is a good reason not to: we do not know how to incorporate it into a theoretical framework within which we can interpret observational data.

Therefore we consider the universe as a 'cosmic fluid' whose 'atoms' are galaxies, How can we get our bearings in such a fluid? We need to set up a coordinate system in which to discuss events. For the time coordinate *t* we take the proper time measured by standard clocks falling freely with the fluid. These clocks lie at the intersections of the grid of spatial coordinate lines. Thus the coordinate grid is expanding with the galaxies, just as grid lines drawn on a rubber balloon expand as it is blown up. These coordinates, which share in the expansion of the universe, are described as '*comoving coordinates*'.

To synchronise the comoving clocks we use the *cosmological principle*. We can state this precisely as follows: '*Every comoving observer in the cosmic fluid has the same history*.' By 'history' here we mean the totality of the observer's experiences and perceptions. Of course, in the actual inhomogeneous universe the histories will not be the same in detail: the rise and decay of civilisations, the evolution and death of particular stars and galaxies, will all be very different in different parts of the universe. But the histories of *averaged* quantities will be the same, if the uniform model has any validity, and it is the values of these quantities that can be employed to synchronise the clocks. For example, all comoving observers could agree (in principle!) to set their clocks at the same time *t* when they see that the mean matter density ρ in their neighbourhood has reached a certain agreed value; alternatively, instead of ρ, the temperature of the microwave background radiation could be used, since (section 8.1) this changes with time as the universe expands.

The cosmological principle essentially states that the large-scale universe is homogeneous. We can also demand that the universe is *isotropic*, that is, that each comoving observer sees the same in all directions in the cosmic fluid. This requirement is very strongly suggested by experiment: the temperature of the microwave background radiation is independent of direction to one part in a thousand, according to a variety of experiments on various scales of angular resolution down to 1'. In addition, the distribution of distant radio and optical galaxies seems to be isotropic. Sometimes isotropy is included in the statement of the cosmological principle.

Taken together, the assumptions of homogeneity and isotropy greatly restrict the geometry of the three-dimensional position space determined by $t = $ const. Obviously this space must be specified by a single curvature, which must be the same at all positions but which may depend on time; we write it as $K(t)$. We have already derived the metric for such a space of constant curvature; it is given by (4.3.8) in terms of polar coordinates r, θ, ϕ, where the r-sphere has defined proper area $4\pi r^2$. However, r is not the radial comoving coordinate we seek, because the area of the r-sphere does not increase as the universe expands. To find the correct coordinate, we first write

▶ $$K(t) \equiv \frac{k}{R^2(t)}, \quad \begin{cases} k = +1 \text{ (positive curvature)} \\ k = 0 \text{ (flat position space)} \\ k = -1 \text{ (negative curvature)} \end{cases} \qquad (6.1.1)$$

where $R(t)$ is called the *cosmic scale factor*. (For positively-curved universes ($k = +1$), there would be some justification in calling $R(t)$ the 'radius of the universe at time t', because (4.3.11) shows that the maximum area of the hyperspheres $r = $ const. is $4\pi R^2(t)$.) Now we define a dimensionless coordinate σ by

▶ $$\sigma \equiv r/R(t), \qquad (6.1.2)$$

so that the spatial metric (4.3.8) becomes

$$\Delta s^2 = R^2(t) \left[\Delta\sigma^2/(1 - k\sigma^2) + \sigma^2 \Delta\theta^2 + \sigma^2 \sin^2\theta \Delta\phi^2 \right].$$

In these coordinates, the proper distance between any two points defined by fixed values of σ, θ, ϕ changes with time in the same way, and this is precisely the property we require of our comoving coordinates in an expanding universe.

To get the complete spacetime separation formula, we simply realise that, along the world line of a comoving observer, where σ, θ and ϕ are constant, coordinate time t must be the same as proper time τ; thus

▶ $$\Delta\tau^2 = \Delta t^2 - R^2(t) \left[\Delta\sigma^2/(1 - k\sigma^2) + \sigma^2 \Delta\theta^2 + \sigma^2 \sin^2\theta \Delta\phi^2 \right]/c^2. \quad (6.1.3)$$

This is called the Robertson–Walker metric, after the relativists who first derived it in 1934. It can be verified that σ, θ, ϕ constant is one of the sets of geodesics in this spacetime, so that the comoving observers really are in free fall. In no sense does our cosmic coordinate system involve a return to absolute space and time; we are simply using the most natural frame of reference, namely the one at rest relative to the averaged local matter of the universe. Finally, we emphasise that no specifically gravitational arguments were employed in the derivation of equation (6.1.3); the Robertson–Walker metric is a consequence only of the symmetry of three-dimensional position space, expressed in a four-dimensional language.

Now let us see how simply the function $R(t)$ describes the expansion of the universe. Imagine for simplicity that our own galaxy, which is approximately comoving, lies at the spatial origin $\sigma = 0$, and consider another galaxy at σ, say. Its proper distance from us at a given cosmic time t is D_p, given by (6.1.3) as

$$\blacktriangleright \quad D_p = R(t) \int_0^\sigma \frac{d\sigma}{\sqrt{(1-k\sigma^2)}} = \begin{cases} R(t)\sin^{-1}\sigma & (k=1) \\ R(t)\,\sigma & (k=0) \\ R(t)\sinh^{-1}\sigma & (k=-1). \end{cases} \quad (6.1.4)$$

Thus the distance is proportional to the cosmic scale factor $R(t)$ which changes with time. The proper velocity v_p is obtained by differentiating with respect to t, realising that σ remains constant because it is a comoving coordinate. Using dots to denote differentiation with respect to time, we obtain

$$\blacktriangleright \quad v_p \equiv \dot{D}_p = \dot{R}(t) \int_0^\sigma \frac{d\sigma}{\sqrt{(1-k\sigma^2)}} = \frac{\dot{R}(t)}{R(t)} D_p. \quad (6.1.5)$$

This tells us that at any given cosmic time t the speed of a galaxy relative to us is proportional to its proper distance from us. But this is simply Hubble's law (2.3.3), with the constant H given by

$$\blacktriangleright \quad H(t) = \frac{\dot{R}(t)}{R(t)}. \quad (6.1.6)$$

We see that in general the expansion rate changes with time; what astronomers measure is the value H_0 of Hubble's constant at the present epoch $t \equiv t_0$.

The case of positive curvature ($k = 1$) is fascinating. We recall from section 4.3 that this describes a *closed universe*, whose three-dimensional space is analogous to the surface of a sphere. As the coordinate ranges from 0 to 1 and back to 0, the 'σ-spheres' sweep out the entire universe. The proper distance at time t from 0 to σ is, from (6.1.4), $R(t)\sin^{-1}\sigma$, which is ambiguous by $2n\pi R(t)$, where n is any integer. This corresponds to measuring the distance to σ with a measuring tape that lies along a spacelike geodesic winding n times round the universe. Although closed, a positively-curved universe is unbounded; as with the surface of a sphere, no 'edge' is ever encountered.

Of course we cannot stretch measuring tapes between galaxies, so proper distance D_p is not a measurable quantity. However, the great advantage of the relativistic formulation is that it gives relationships between quantities such as red shifts, apparent magnitudes, number counts etc., which can be measured. In the next few sections we shall derive these relation-

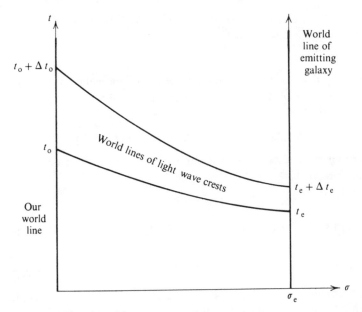

Figure 50. World lines for calculating cosmic red shifts.

ships; the important point is that they all involve only the scale factor $R(t)$.

How can $R(t)$ be determined theoretically? The answer to this requires a cosmic dynamics, in the form either of a gravitational theory such as Einstein's or of a steady state hypothesis. We shall learn in chapter 7 that a variety of model universes can be described in this way, characterised by different scale factors $R(t)$. It is the task of observational cosmology to find out which of these models (if any) describes our actual universe.

6.2 Red shifts and horizons

Consider light reaching us (at $\sigma = 0$) at the present time t_0 from a distant galaxy at $\sigma = \sigma_e$ (figure 50). Two crests arriving at t_0 and $t_0 + \Delta t_0$ were emitted at t_e and $t_e + \Delta t_e$. Since the light has travelled radially inwards along a null geodesic, the Robertson–Walker metric (6.1.3) can be employed to relate these four times to σ_e. Since

$$0 = \Delta t^2 - R^2(t)\Delta\sigma^2/c^2(1 - k\sigma^2)$$

we have

$$\int_{t_e}^{t_0} \frac{dt}{R(t)} = \frac{1}{c}\int_0^{\sigma_e} \frac{d\sigma}{\sqrt{(1 - k\sigma^2)}} \quad \text{and} \quad \int_{t_e+\Delta t_e}^{t_0+\Delta t_0} \frac{dt}{R(t)} = \frac{1}{c}\int_0^{\sigma_e} \frac{d\sigma}{\sqrt{(1 - k\sigma^2)}},$$

so that

$$\int_{t_e+\Delta t_e}^{t_o+\Delta t_o} \frac{dt}{R(t)} - \int_{t_e}^{t_o} \frac{dt}{R(t)} = 0.$$

For all kinds of electromagnetic radiation received from galaxies, Δt_e and Δt_o are tiny fractions of a second, and over such small times $R(t)$ hardly changes; therefore

$$\Delta t_o/R(t_o) - \Delta t_e/R(t_e) = 0, \quad \text{or} \quad \Delta t_o/\Delta t_e = R(t_o)/R(t_e).$$

Now, the observed and emitted wavelengths λ_o and λ_e are related to the periods Δt_o and Δt_e by

$$\lambda_o = c\Delta t_o \quad \text{and} \quad \lambda_e = c\Delta t_e$$

so that the red shift z, defined by equation (2.3.1), can be written in terms of $R(t_o)$ as

▶ $\quad z = R(t_o)/R(t_e) - 1.$ (6.2.1)

In an expanding universe, $R(t_o) > R(t_e)$, so that z is positive, as observed. Formula (6.2.1) shows that the red shift gives a direct measure of the degree to which the universe has expanded since the time t_e when the radiation was emitted (of course t_e is not known, but we shall soon express other observable quantities in terms of $R(t)$, and so relate them to z).

At this point we introduce a useful mathematical expansion. Most observed cosmological red shifts are rather small, so that t_e is (cosmologically speaking) not much earlier than t_o. Therefore we expand $R(t_e)$ about t_o:

$$R(t_e) = R(t_o) + (t_e - t_o) \dot{R}(t_o) + \tfrac{1}{2}(t_e - t_o)^2 \ddot{R}(t_o) + \dots$$
$$\equiv R(t_o)[1 + H_0(t_e - t_o) - \tfrac{1}{2}q_0 H_0^2(t_e - t_o)^2 + \dots],$$

where H_0 is the present value of Hubble's constant, i.e.

▶ $\quad H_0 \equiv \dot{R}(t_o)/R(t_o),$ (6.2.2)

and q_0 is the dimensionless *deceleration parameter*

▶ $\quad q_0 \equiv -\ddot{R}(t_o) R(t_o)/\dot{R}^2(t_o) = -\ddot{R}(t_o)/R(t_o) H_0^2.$ (6.2.3)

(Clearly q_0 is positive if the expansion of the universe is slowing down, i.e. if $\ddot{R}$ is negative. As we shall see in chapter 7, this is the case for most models of the universe, so it is convenient to define a deceleration parameter rather than an acceleration parameter.) The red shift (6.2.1) may now be expanded in powers of $t_o - t_e$, as

▶ $\quad z = H_0(t_o - t_e) + (1 + \tfrac{1}{2}q_0) H_0^2(t_o - t_e)^2 + \dots,$ (6.2.4)

and, conversely, the time of light travel $t_0 - t_e$ may be expanded as a function of z:

▶ $$t_0 - t_e = \frac{1}{H_0} [z - (1 + \tfrac{1}{2} q_0) z^2 + \ldots]. \tag{6.2.5}$$

These formulae will be very useful later, but it should not be forgotten that they are only approximations, valid for small z. Meanwhile, we anticipate the results of chapter 7, and state that H_0, q_0 and the present mass density ρ_0 of the cosmic fluid, together with the laws of cosmic dynamics, suffice to determine completely the entire form of the scale factor $R(t)$.

Now we ask: what is the coordinate σ_{oh} of the most distant object (e.g. a galaxy) we can see now? This coordinate is called the *object horizon*, or the *particle horizon*. Such an object must have emitted its light at the beginning of the universe t_{min}; on some models t_{min} is $-\infty$, on others t_{min} is zero. Therefore, we have, from the Robertson–Walker metric,

$$\int_{t_{min}}^{t_0} \frac{dt}{R(t)} = \frac{1}{c} \int_0^{\sigma_{oh}} \frac{d\sigma}{\sqrt{(1 - k\sigma^2)}},$$

or (cf. (6.1.4))

▶ $$\sigma_{oh} = \begin{cases} \sin \left(c \int_{t_{min}}^{t_0} \frac{dt}{R(t)} \right) & (k = 1) \\[2ex] c \int_{t_{min}}^{t_0} \frac{dt}{R(t)} & (k = 0) \\[2ex] \sinh \left(c \int_{t_{min}}^{t_0} \frac{dt}{R(t)} \right) & (k = -1). \end{cases} \tag{6.2.6}$$

Objects beyond σ_{oh} cannot now be seen by us (the light is like a runner in a race where the winning post recedes faster than he can run). If the space is flat or negatively curved, it is possible that, for certain forms of the function $R(t)$, σ_{oh} is infinite; such a model universe has no object horizon, and in principle the whole universe could be seen by us now. For a positively-curved universe there is no object horizon if

$$c \int_{t_{min}}^{t_0} \frac{dt}{R(t)} \geqslant \pi;$$

then the observable objects can have values of σ ranging from 0 to 1 and back to 0, and we see the 'antipodes' and hence the entire universe (imagine a beetle setting out from the south pole of an expanding balloon trying to crawl to an observer sitting at the north pole). The introduction of this idea of an object horizon renders precise the concept of 'the limit of the observable universe' that we discussed in section 2.3 as a consequence of Hubble's law.

No physical influence can travel faster than light. Therefore the object horizon represents the greatest distance from which outside matter could have affected what is now happening at any given locality. In the early epochs of 'big bang' model universe, the object horizons were very small; later, the horizons increased and mutual influence between masses became possible within ever larger comoving volumes. It is possible that this may help to explain how the present large-scale homogeneity and isotropy of the universe evolved from an initial 'chaos' containing perturbations of all sizes (see section 8.2).

Now we ask a different question: what is the coordinate σ_{eh} of the most distant event occurring now (that is, at cosmic time t_0) that we shall ever be able to see? This coordinate is called the *event horizon*. The light from such an event must reach us before the universe ends, at t_{max}; on most models, t_{max} is infinite, but there are important models (see chapter 7) where the expansion stops and is replaced by a contraction which ends in a gigantic implosion at a finite time t_{max}. Therefore we have

$$\int_{t_0}^{t_{max}} \frac{dt}{R(t)} = \frac{1}{c} \int_0^{\sigma_{eh}} \frac{d\sigma}{\sqrt{(1 - k\sigma^2)}},$$

or (cf. (6.1.4))

$$\sigma_{eh} = \begin{cases} \sin\left(c \int_{t_0}^{t_{max}} \frac{dt}{R(t)}\right) & (k = 1) \\ c \int_{t_0}^{t_{max}} \frac{dt}{R(t)} & (k = 0) \\ \sinh\left(c \int_{t_0}^{t_{max}} \frac{dt}{R(t)}\right) & (k = -1). \end{cases} \qquad (6.2.7)$$

Events now occurring beyond σ_{eh} will never be seen by us. For negatively-curved or flat universes in which σ_{eh} is infinite, or for positively-curved universes in which

$$c \int_{t_0}^{t_{max}} \frac{dt}{R(t)} \geq \pi,$$

any event happening now could sometime be seen by us.

The event horizon represents the greatest distance from which outside matter could eventually affect what will happen at any given locality. An example of an event horizon is the Schwarzschild sphere $r = r_S$ discussed in section 5.6.

Later, when we have worked out $R(t)$ for various cosmological models, we shall know which models of the universe possess horizons of either type, and which do not.

6.3 Apparent luminosity

We recall the discussion in section 2.2.3 of the apparent luminosity l of an object (which will be a galaxy or a radio source in all that follows) whose absolute luminosity is L and whose distance is D_L ; we rewrite the simple formula (2.2.5) as

▶ $$l = L/4\pi D_L^2. \tag{6.3.1}$$

The derivation was based on the conservation of energy, and took no account of the expansion of the universe or the curvature of space. Within the uniform universe model, based on the Robertson–Walker metric, it is possible to calculate l exactly. Let the galaxy be at cosmic radial coordinate σ_e, and consider the light that it emitted at cosmic time t_e, which is just reaching us now, at t_0. This light is now crossing the surface of a large hypersphere whose area, from (6.1.2), is

$$4\pi R^2(t_0)\,\sigma_e^2.$$

The power l crossing unit area of this sphere is reduced by two effects in addition to the inverse-square law: first, because of the red shift, each photon arrives with its energy reduced by a factor $1+z$ (remember that the energy of a photon is $h\nu$, where h is Planck's constant and ν is the frequency); second, applying to photons the same arguments that for wave crests led to the red shift, we see that the rate of reception of photons is less than their rate of emission, also by a factor $1+z$. Therefore the apparent luminosity is

▶ $$l = L/4\pi R^2(t_0)\,\sigma_e^2(1+z)^2. \tag{6.3.2}$$

Combining this with the formula (6.2.1) for red shift gives

▶ $$l = LR^2(t_e)/4\pi\sigma_e^2 R^4(t_0). \tag{6.3.2a}$$

We can express σ_e in terms of t_0 and t_e, because the light travels along a null geodesic, so that:

▶ $$l = LkR^2(t_e)/4\pi R^4(t_0)\sin^2\left(c\sqrt{k}\int_{t_e}^{t_0} dt/R(t)\right) \tag{6.3.3}$$

(cf. the beginning of section 6.2 and equation (6.1.4)). For an object of known absolute luminosity L, application of the simple law (6.3.1) to a measurement of l yields the *luminosity distance* D_L, which is what the methods of section 2.2.3 really measure. The exact result (6.3.3) tells us that

$$\blacktriangleright \qquad D_L = R^2(t_0) \sin\left(c\sqrt{k}\int_{t_e}^{t} dt'/R(t)\right) \bigg/ R(t_e)\sqrt{k}, \qquad (6.3.4)$$

or

$$\blacktriangleright \qquad D_L = R^2(t_0)\, \sigma_e/R(t_e). \qquad (6.3.5)$$

This is not the same as the *proper distance* D_P, given by (6.1.4). Nor is it the same as the *coordinate distance* $R(t_0)\, \sigma_e$, or the *distance inferred from apparent size* or the *distance inferred from parallax*. However, all these distances become the same when σ_e is small, that is, when $t_0 - t_e$ is small.

For galaxies that are not too distant, it is sensible to expand (6.3.3) in powers of $t_0 - t_e$; we obtain, after a little reduction,

$$l = \frac{L}{4\pi c^2(t_0 - t_e)^2}\left[1 - \frac{3\dot{R}(t_0)}{R(t_0)}(t_0 - t_e) + \ldots\right].$$

Now we use the expansion (6.2.5) for $t_0 - t_e$ in terms of the red shift z:

$$l = \frac{LH_0^2}{4\pi c^2 z^2}\,[1 + (q_0 - 1)\, z + \ldots]. \qquad (6.3.6)$$

The beauty of this result is that it relates directly the measurable quantities l and z. We do not know L with any great accuracy, but we can estimate it from the cosmic distance hierarchy as explained in section 2.2. Knowing L for galaxies close enough to us for $(q_0 - 1)\, z$ to be negligible, we can find H_0 from the slope of the curve of l v. z^{-2}. Then, assuming that the brightest galaxies in clusters all have the same L, we can extend the method out to greater distances, and obtain q_0 from the deviation of the curve of l v. z^{-2} from a straight line. Actually it is more common in optical astronomy to express l and L in terms of apparent and absolute magnitudes m and M using (2.2.6); the luminosity formula (6.3.6) now becomes, after some reduction,

$$\blacktriangleright \qquad m - M = 25 - 5\log_{10} H_0 + 5\log_{10} cz + 1.086(1 - q_0)\, z \qquad (6.3.7)$$

where H_0 is in km s^{-1} Mpc^{-1} and c is in km s^{-1}.

Attempts to use this formula to find H_0 and q_0 are beset with technical complications. The most basic problem is lack of statistics for red shifts $z \gtrsim 0.3$. Then there is the assumed constancy of L or M: apart from the Scott effect mentioned in section 2.2.3, concerning the artificial selection of extraordinarily bright galaxies, there is the question of *galactic evolution*. The light from a galaxy with a large red shift was emitted at a time t_e when that galaxy was much younger than it is at the present cosmic time;

therefore it might have had a very different absolute luminosity from present-day galaxies. Computer models of an evolving galaxy can give some idea of how the luminosity changes, and make possible a very rough correction to the measured magnitudes m. There are also a number of other corrections, for example for absorption within our galaxy, which affects m, and rotation of our galaxy, which affects z. After all corrections have been made, the best fit to the z v. m curve (figure 6) gives the value (55 ± 7) km s^{-1} Mpc^{-1} for H_0 and for q_0, the deceleration parameter,

▶ $\qquad q_0 = 1 \pm 1.$ (6.3.8)

(Some recent indirect arguments suggest a smaller value for q_0, such as 0.2 or 0.3.) The uncertainties are so great that, to quote Sandage, 'At present none of this can be taken very seriously. We need many more clusters whose red shifts are greater than $z = 0.2$ for a satisfactory solution.' However, one important tentative conclusion can be drawn from the observations: it is rather unlikely that the steady state model applies to the actual universe because, as we shall show in chapter 7, this model requires q_0 to be -1.

6.4 Galactic densities and the darkness of the night sky

Let us assume for simplicity that the galaxies which make up the cosmic fluid at any given cosmic time all have the same average luminosity, $L(t)$; this may change with time, so we are now allowing for the possiblity of galactic evolution. Let the number of 'large galaxies' in unit proper volume be $n(t)$; this is the 'proper number density'. The number of galaxies enclosed between the coordinate hyperspheres σ and $\sigma + d\sigma$ at t is, from the Robertson–Walker metric (6.1.3),

▶ $\qquad n(t) \times (\text{proper volume between } \sigma \text{ and } \sigma + d\sigma \text{ at } t)$

$\qquad = n(t) \underbrace{4\pi\sigma^2 R^2(t)}_{\text{proper area}} . R(t) \, d\sigma / \sqrt{(1 - k\sigma^2)}$

$\qquad = 4\pi n(t) \, R^3(t) \, \sigma^2 \, d\sigma / \sqrt{(1 - k\sigma^2)}.$ (6.4.1)

If galaxies are neither created nor destroyed, this number remains constant, because σ and $\sigma + d\sigma$ are comoving spheres, so that galaxies do not cross their surfaces, on the average. Thus

▶ $\qquad n(t) \, R^3(t) = n(t_0) \, R^3(t_0),$

or $\qquad n(t) = \dfrac{R^3(t_0)}{R^3(t)} \, n(t_0).$ (6.4.2)

This formula is really obvious: volumes defined by galaxies expand as the cube of the scale factor, so that densities decrease as $R^{-3}(t)$. (On the steady state model (6.4.2) does not hold, and galaxies are assumed to be created at such a rate that $n(t)$ remains constant as the universe expands – see section 7.3.) The present density $n(t_0)$ of galaxies is estimated to be $1/75$ Mpc^{-3}.

The notion of density, and the results of the last section on luminosity, enable us to formulate a condition which cosmological models must satisfy in order not to predict a night sky ablaze with light, that is, a condition for the avoidance of 'Olbers' paradox'. Obviously, the total apparent luminosity l_{tot}, due to all the galaxies in the universe, must not be infinite. Now we calculate an upper limit for l_{tot}, making use of (6.4.1) and the luminosity law (6.3.2b) and neglecting the absorption effects arising from the obscuration of one galaxy by another. The apparent luminosity dl_{tot} at the present time t_0 from all the galaxies between σ_e and $\sigma_e + d\sigma_e$ is

$$dl_{tot} = \text{No. of galaxies} \times \text{apparent luminosity of each}$$

$$= \frac{4\pi n(t_e) R^3(t_e) \sigma_e^2 d\sigma_e}{\sqrt{(1 - k\sigma_e^2)}} \times \frac{L(t_e) R^2(t_e)}{4\pi \sigma_e^2 R^4(t_0)},$$

where t_e is the time of emission from σ_e of light reaching us now. t_e and t_0 are related by the geodesic law for light tracks:

$$\blacktriangleright \qquad \frac{c\,dt_e}{R(t_e)} = \frac{d\sigma_e}{\sqrt{(1 - k\sigma_e^2)}}. \tag{6.4.3}$$

Thus we can express dl_{tot} in terms of dt_e

$$dl_{tot} = \frac{cn(t_e) L(t_e) R^4(t_e)\,dt_e}{R^4(t_0)}.$$

To find l_{tot}, we integrate over the entire past history of the universe, from its origin t_{min} up to the present; this gives

$$\blacktriangleright \qquad l_{tot} = \frac{c}{R^4(t_0)} \int_{t_{min}}^{t_0} dt_e\, n(t_e) L(t_e) R^4(t_e). \tag{6.4.4}$$

To prevent a blazing night sky, this quantity must not be infinite. If galaxies are conserved, we may eliminate $n(t_e)$ using (6.4.2), and l_{tot} becomes

$$\blacktriangleright \qquad l_{tot} = \frac{cn(t_0)}{R(t_0)} \int_{t_{min}}^{t_0} dt_e\, L(t_e) R(t_e). \tag{6.4.5}$$

For 'big bang' models with $t_{min} = 0$, $R(t_{min})$ is zero by definition, and $L(t_e)$ is always finite; thus the integral converges, and l_{tot} is finite. This is not a sufficient condition for the avoidance of Olbers' problem, because the actual value calculated for l_{tot} (corrected for absorption) must be less than the known intensity of the radiation background of every wavelength;

this limitation on l_{tot} is most severe in the X-ray and radio regions of the spectrum.

For infinitely old universes, where $t_{min} = -\infty$, the possibility arises that l_{tot} might be infinite, and we must have, for (6.4.5) to converge,

▶ $L(t) R(t) < \text{const.}/|t| \quad \text{as} \quad t \to -\infty.$ (6.4.6)

This is violated by the static models current before the expansion of the universe was discovered; n, L and R are constant in these models, so that

$$l_{tot} = cnL \int_{-\infty}^{t_0} dt_e = \infty.$$

(Absorption will reduce these 'infinite' calculated luminosities to a value corresponding to the whole sky being covered with stars, but Olbers' problem remains.)

6.5   Number counts

How many of the galaxies we now see emitted their light after t_e? From (6.4.1) and (6.4.3), this number, which we call $N(> t_e)$, is given by

▶ $$N(> t_e) = \int_{t_e}^{t_0} dt \frac{4\pi n(t) R^3(t) \sigma^2(t) d\sigma}{\sqrt{(1-k\sigma^2)} dt}$$

$$= 4\pi c \int_{t_e}^{t_0} \frac{dt\, n(t) R^2(t)}{k} \sin^2\left(c\sqrt{k} \int_t^{t_0} \frac{dt'}{R(t')}\right).$$ (6.5.1)

For standard galaxy-conserving cosmologies, (6.4.2) gives

▶ $$N(> t_e) = \frac{4\pi c R^3(t_0) n(t_0)}{k} \int_{t_e}^{t_0} \frac{dt}{R(t)} \sin^2\left(c\sqrt{k} \int_t^{t_0} \frac{dt'}{R(t')}\right).$$ (6.5.2)

Now, $N(> t_e)$ is also the number $N(< z)$ of galaxies with red shifts less than z, where z is the red shift corresponding to t_e, given by (6.2.1), because galaxies emitting *since* t_e have red shifts *less than* z. Furthermore, $N(> t_e)$ is also the number $N(> l)$ of galaxies with apparent luminosities greater than l, where l is the luminosity corresponding to t_e, given by (6.3.3), because galaxies emitting since t_e appear brighter than l. The functions $N(< z)$ and $N(> l)$ are two relationships between pairs of observable quantities. Unfortunately these relations are implicit in their general forms, and all we can obtain explicitly are power series expansions for small z or large l.

We begin by expanding $N(> t_e)$ (equation (6.5.2)) for small $t_0 - t_e$; this gives

$$N(> t_e) = \frac{4\pi c^3 n(t_0)}{3}(t_0 - t_e)^3 \left[1 + \frac{3H_0}{2}(t_0 - t_e) + \ldots\right].$$

Now we employ the red shift expansion (6.2.5) to express $t_0 - t_e$ in terms of z, to obtain

$$\blacktriangleright' \qquad N(< z) = \frac{4\pi c^3 n(t_0) z^3}{3H_0^3} [1 - \tfrac{3}{2}(1 + q_0) z + ...]. \qquad (6.5.3)$$

This very much resembles the l v. z relation (6.3.6). In principle it can be used to determine H_0 and the deceleration parameter q_0; the values found should of course be the same as those found previously, and this would provide a most valuable check on the validity of uniform cosmologies based on the Robertson–Walker metric. Unfortunately, this programme has not yet been carried out, for the following reasons. In the case of galaxies visible in optical telescopes, it is easy to measure z, but attempts to obtain reliable number counts run into great technical difficulties, principally arising from clustering, and obscuring matter within our own galaxy. In the case of radio sources, on the other hand, counting is possible, but no lines arising from sources are seen in the radio spectrum, so that z can be determined only in those cases where the source can be identified optically. A further complication arises because galaxies and radio sources do not all have the same luminosity, as we have assumed, nor do the luminosities of individual objects remain constant.

To find the number–luminosity relation $N(> l)$, we substitute into (6.5.3) the value of z in terms of l given by (6.3.6). This gives

$$\blacktriangleright \qquad N(> l) = \frac{4\pi n(t_0)}{3} \left(\frac{L}{4\pi l}\right)^{\frac{3}{2}} \left[1 - \frac{3H_0}{c}\left(\frac{L}{4\pi l}\right)^{\frac{1}{2}} + ...\right]. \qquad (6.5.4)$$

The leading term of this formula is the Euclidean $l^{-\frac{3}{2}}$ law (2.2.12); it enables $n(t_0)$ to be found, as discussed in section 2.2.4. The correction term (which, by a fortuitous cancellation, does not involve q_0) is always negative, so that for faint objects (l small), there should always be *fewer* sources than the $l^{-\frac{3}{2}}$ law predicts. This conclusion is flatly contradicted by observations on radio sources: many radio surveys, involving thousands of sources, all agree that there are *more* faint sources than the $l^{-\frac{3}{2}}$ law predicts. The experimental results are well fitted by a law of the form

$$\blacktriangleright \qquad N(> l) \approx \text{const.}/l^{1.8}. \qquad (6.5.5)$$

What are we to make of this? Obviously one or more of the approximations leading to (6.5.4) must be inapplicable. These approximations are that galaxies are conserved and that their average absolute luminosity L remains constant. Since the formula is contradicted for small l, which corresponds to faint distant sources whose radiation was emitted at an early time t_e in the history of the universe, we must conclude that *in the past the radio sources were brighter and/or more numerous than they are today.* Thus

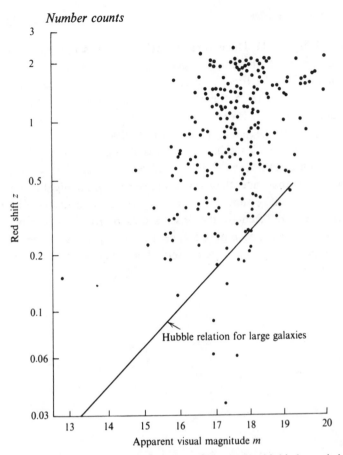

Figure 51. Red shifts of QSOs. (Reproduced with kind permission of Professor E. M. Burbidge.)

evolution must be taken into account in interpreting observations (of course, radio and optical telescopes measure completely different fractions of the total luminosity L; this complicates the details of the argument without affecting its validity).

Experiment tells us more than this, however. The $l^{-1.8}$ law, which gives the first correction to $l^{-\frac{3}{2}}$ as l is reduced, cannot hold for very small values of l, for the following reason: the very faintest sources cannot be resolved, and their radio emission forms a continuous background; if the $l^{-1.8}$ law held for these unresolved sources, this background would be much more intense than it actually is. In fact for the faintest resolvable sources the $l^{-1.8}$ law does seem to change to a gentler dependence – roughly $l^{-0.8}$. This means that at an earlier epoch than that to which (6.5.5) applies, the radio sources must have been fainter and/or less numerous than today. Observations on QSOs support this conclusion: their red shifts rarely exceed $z = 2.2$ (figure 51). The nature of QSOs is still obscure, but if they are extragalactic objects this result implies that they did not emit significant radiation before the time t_e corresponding to $z = 2.2$ (the precise value of t_e depends on $R(t)$, and hence on the cosmological model).

All this points to the following simple scenario for the history of the universe after the very beginning (which will be discussed in chapter 8): at first there were no significant discrete sources of radiation, then galaxies, radio sources and QSO's began to form, and radiated abundantly during their infancy. Since that time, these objects have been gradually dimming. In section 7.3. we shall see that the contrasting picture, of a universe in a steady state, is contradicted by the observations of $N(> l)$ (we cannot make this assertion on the basis of (6.5.4) because that formula was derived on the assumption that galaxies are conserved, and this is violated on the steady state theory).

7 *Cosmic dynamics*

7.1 Gravitation and the cosmic fluid

The uniform model universes based on the Robertson–Walker separation formula (6.1.3) are characterised by their scale factors $R(t)$ and 'curvature index' k. According to (6.1.1), these quantities determine the curvature $K(t)$ of three-dimensional position space at cosmic time t. What determines $R(t)$ and k? The answer is: the self-gravitation of all the matter in the universe, since this must control the nature of the expansion. In a rigorous treatment, the curvature tensor of spacetime would be given, by Einstein's field equations, in terms of the tensor describing the distribution of cosmic matter. However, such generality involves mathematics too advanced for the present work, and we find the connection between curvature and matter by using arguments similar to those of section 4.4.

First we realise that it is not possible to use formula (4.4.1), which gives the spatial curvature near an isolated mass. The reason is that this leads to the *static* Schwarzschild spacetime (5.1.2), whereas an expanding universe is certainly not static. Nevertheless, we can use similar arguments, provided we apply them to the curvature $\mathcal{K}(t)$ of *spacetime* rather than to the curvature of space. $\mathcal{K}(t)$ is defined as the curvature of the two-dimensional 'surface' specified by varying t and σ, keeping θ and ϕ constant. The separation formula is, from (6.1.3):

$$\Delta\tau^2 = \Delta t^2 - R^2(t)\,\Delta\sigma^2/c^2(1-k\sigma^2).$$

To find $\mathcal{K}(t)$ we use the Gauss curvature formula (4.3.5), with $t = x^1$, $\sigma = x^2$. This leads to

▶ $$\mathcal{K}(t) = -\ddot{R}(t)/R(t). \tag{7.1.1}$$

The matter in the universe is described by its mass density $\rho(t)$, and is the cause of the curvature of spacetime. In the 'simplest possible' general formula, $\mathscr{K}(t)$ is proportional to $\rho(t)$, and we write

▶ $\qquad \mathscr{K}(t) = \alpha\rho(t)\, G^l c^m + \text{const.},$ $\qquad\qquad$ (7.1.2)

where the 'constant' is inserted to allow for the possibility that empty spacetime ($\rho = 0$) might be curved. Now we use dimensional analysis: from (7.1.1) it follows that $\mathscr{K}(t)$ has dimensions (time)$^{-2}$. If α is dimensionless, this implies that the powers l and m to which G and c are raised are 1 and 0 respectively. Soon we shall see that the correct Newtonian limit requires us to take

$\qquad \alpha = 4\pi/3.$

Finally, to agree with the usual convention we write the 'constant' in (7.1.2) as $-\Lambda/3$; Λ has dimensions (time)$^{-2}$ and is called the *cosmical constant*. The curvature $\mathscr{K}(t)$ now becomes

▶ $\qquad \mathscr{K}(t) = 4\pi\rho(t)\, G/3 - \Lambda/3.$ $\qquad\qquad$ (7.1.3)

Equating this with (7.1.1) gives, as the equation of motion for $R(t)$,

▶ $\qquad \ddot{R}(t) = -4\pi\rho(t)\, GR(t)/3 + (\Lambda/3)\, R(t).$ $\qquad\qquad$ (7.1.4)

Precisely the same formula follows from Einstein's general field equations. The cosmical constant Λ has been the object of a lot of controversy. It is a new independent constant of nature, like G and c, and should surely be avoided if possible. Yet the form (7.1.2), *with* the constant, is more general, and it does seem unlikely that Λ is exactly zero. Einstein himself believed very strongly that Λ is zero, and he later referred to the original inclusion of a cosmical constant in his field equations as the 'biggest blunder of my life'. His argument was that a non-zero Λ requires empty spacetime to be curved, and this is contrary to the spirit of Mach's principle (section 3.5). We shall compromise, retaining Λ but reserving the right to set it equal to zero when it suits us. From (7.1.4) it is clear that a positive Λ acts like a negative density ρ. Since the self-gravitation of matter acts to slow down the expansion of the universe, a positive Λ must act to accelerate it; for this reason $\Lambda R/3$ is sometimes called the *cosmic repulsion term*.

It helps in the understanding of (7.1.4) to realise that if Λ is zero this equation follows exactly from Newtonian mechanics. To see this, consider the galaxies lying inside the comoving sphere with radial coordinate σ. From (6.1.4), σ corresponds in Newtonian flat space to a proper radius $\sigma R(t)$. A galaxy on the surface of the sphere will accelerate inwards under the attraction of the mass within the sphere. If the galaxy has mass m, Newton's second law gives

$$m \times \text{acceleration} = m\sigma \ddot{R}(t) = \text{outward force on galaxy}$$

$$= -\frac{Gm \times \text{mass inside sphere}}{(\sigma R(t))^2}$$

$$= -\frac{Gm(4\pi/3)\,\rho(t)\,(\sigma R(t))^3}{(\sigma R(t))^2}.$$

Thus

$$\ddot{R}(t) = -(4\pi/3)\,G\rho(t)\,R(t),$$

which is identical with (7.1.4) if $\Lambda = 0$. This shows that our choice of $\alpha = 4\pi/3$ in (7.1.3) was necessary in order to obtain compatibility with Newtonian mechanics. Compatibility is one thing, but identity is another; why does Newtonian mechanics give the exact answer to this problem? The reason is that the motion of the whole cosmic fluid is determined by the motion of any very small sphere of fluid, according to the cosmological principle. Relative expansion speeds within such a small sphere are very much less than c, and gravitational forces within the sphere are very small because the mass in the sphere is small. However, it is precisely in the limit of low speeds and weak gravitation that Newtonian mechanics is valid. Thus the Newtonian equation governs the motion of the whole cosmic fluid. (Even the cosmical constant Λ can be incorporated into the Newtonian scheme by postulating a long-range repulsive force proportional to distance, acting in addition to gravity.) However, it must not be imagined that the applicability of Newtonian dynamics means that general relativity is wholly unnecessary in cosmology, because the formulae of chapter 6 for red shifts, luminosities and number counts follow from the kinematics of light signals, and these essentially relativistic results inescapably involve non-Euclidean geometry.

As a first step towards solving (7.1.4) we eliminate $\rho(t)$ using the law of conservation of matter: within a comoving sphere the mass remains constant, while the volume is proportional to $R^3(t)$; thus

$$\rho(t)\,R^3(t) = \text{const.},$$

or

$$\blacktriangleright \qquad \rho(t) = \frac{\rho(t_0)\,R^3(t_0)}{R^3(t)} \qquad\qquad (7.1.5)$$

(cf. (6.4.2) for the *number* density of galaxies). Now (7.1.4) becomes

$$\ddot{R}(t) = -\frac{4\pi\rho(t_0)\,R^3(t_0)\,G}{3R^2(t)} + \frac{\Lambda R(t)}{3}.$$

Multiplying by $2\dot{R}(t)$ and integrating, we obtain

$$\dot{R}^2(t) = \frac{8\pi\rho(t_0)\,R^3(t_0)\,G}{3R(t)} + \frac{\Lambda}{3}\,R^2(t) + \text{const.}$$

$$= \frac{8\pi}{3}\,G\rho(t)\,R^2(t) + \frac{\Lambda}{3}\,R^2 + \text{const.}$$

The 'constant' must be proportional to c^2, since this is the only combination of the quantities c, G and Λ which has the required dimensions of (velocity)^2. The field equations of general relativity give its exact value as $-kc^2$, where k is the curvature index of equation (6.1.1). Thus we have

▶ $\qquad \dot{R}^2(t) = -kc^2 + (8\pi G\rho(t) + \Lambda)\,R^2(t)/3.$ (7.1.6)

In section 7.2 we shall solve this equation and obtain the form of $R(t)$. First, however, we apply (7.1.4) and (7.1.6) (which hold for all times) to the present cosmic time t_0. Then $\dot{R}(t_0)$ and $\ddot{R}(t_0)$ can be expressed in terms of Hubble's constant H_0 and the deceleration parameter q_0, defined by equations (6.2.2) and (6.2.3). If we denote the present mass density $\rho(t_0)$ by ρ_0, equations (7.1.4) and (7.1.6) become

▶ $\qquad \Lambda = 4\pi\rho_0 G - 3q_0 H_0^2,$ (7.1.7)

and

▶ $\qquad k = (R^2(t_0)/c^2)\,[4\pi G\rho_0 - H_0^2(q_0 + 1)].$ (7.1.8)

The quantities ρ_0, q_0 and H_0 are measurable, so that Λ can be calculated from (7.1.7). The quantity in square brackets in (7.1.8) can also be evaluated; if it is positive, then $k = +1$, and if it is negative, then $k = -1$. Knowing k, (7.1.8) can be used to calculate the present 'radius of curvature' $R(t_0)$ of the universe.

Let us calculate Λ, k and $R(t_0)$ using the current 'best' values. For H_0 we take the value (2.3.3), for q_0 we take (6.3.8), while for ρ_0 we take the mass density ρ_{gal} due to galaxies, given by (2.1.2); the values for ρ_0 and q_0 are grossly uncertain. We obtain

▶ $\qquad \Lambda = (-0.91 \pm 0.93) \times 10^{-20}$ years^{-2},

$\qquad k = -1,$

$\qquad R(t_0) > 10^{10}$ light years $\approx 3 \times 10^9$ pc. (7.1.9)

The negative value of Λ indicates a *cosmic attraction*. On relatively small scales, such as that of the solar system, this is very weak, and the appropriate modifications to Newtonian mechanics lead to effects that could not possibly be detected. The negative value of k indicates an *open* universe with infinite proper volume. However, we emphasise again the great uncertainties in the numerical results: because of the large errors in q_0 it is just possible that k and Λ might both be positive.

An alternative approach is to take the 'pure Einstein' view that the cosmical constant is zero. Then (7.1.7) indicates that the density ρ_0 is no longer an independent quantity, but is given by

$$\stackrel{(\Lambda=0)}{\rho_0} = 3q_0 H_0^2/4\pi G = 1.1 \times 10^{-26} \text{ kg m}^{-3} \approx 40\,\rho_{\text{gal}}, \qquad (7.1.10)$$

where we have again used the current 'best' values for H_0 and q_0. This great difference between ρ_0 and ρ_{gal}, together with the widespread feeling that Λ 'ought' to be zero, has led to the view that there might be many times more matter between galaxies than in them. However, all attempts to find this 'missing mass' have so far been unsuccessful. Assuming that this mass does exist, and therefore that Λ is zero, we can find k and $R^2(t_0)$ from (7.1.8), since

$$\stackrel{(\Lambda=0)}{k/R^2(t_0)} = (H_0^2/c^2)\,(2q_0-1); \qquad (7.1.11)$$

this gives

$$k = +1,$$

$$R(t_0) > 1.05 \times 10^{10} \text{ light-years} \approx 3.2 \times 10^9 \text{ pc}. \qquad (7.1.12)$$

Therefore the present data suggest a *closed* universe if the cosmical constant is zero, although k would be negative, and the universe *open*, if q_0 were less than 0.5, which is certainly not ruled out if $q_0 = 1 \pm 1$. If q_0 were as small as 0.025 there would be no 'missing mass' problem, because (7.1.10) would predict $\rho_0 \approx \rho_{\text{gal}}$.

Now we can see why efforts to get more precise measurements of H_0, q_0 and ρ_0 form such a large part of observational cosmology. Present data do not even tell us for certain whether the universe is open or closed, or whether there is a cosmic attraction or repulsion. In the exploration of model universes which follows, most of our numerical calculations will involve uncertainties of 100 per cent or more, but since these calculations are merely illustrative we shall omit all error bounds.

7.2 Histories of model universes

The cosmic scale factor $R(t)$ provides a complete description of the uniform model universes derived from general relativity and the cosmological principle. To find $R(t)$, we can use the gravitational equation (7.1.6), together with the conservation condition (7.1.5). These give

$$\dot{R}^2(t) = 8\pi G\rho_0\,R^3(t_0)/3R(t) - kc^2 + (\Lambda/3)\,R^2(t)$$

$$\equiv A(R(t)), \text{ say}. \qquad (7.2.1)$$

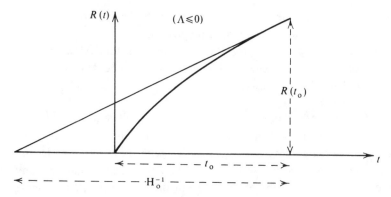

Figure 52. Universe expanding from 'big bang'.

This is Friedmann's equation, first derived and systematically examined in 1922. Now Λ, ρ_0, $R(t_0)$ and k can in principle be found from observations made at the present time t_0, as explained in the last section. Thus the function $A(R)$ is completely known, and we can obtain an implicit equation for $R(t)$ as follows:

$$\frac{dR}{dt} = \sqrt{(A(R))}, \; dt = \frac{dR}{\sqrt{(A(R))}},$$

so that

$$\blacktriangleright \quad t - t_1 = \int_{R(t_1)}^{R(t)} \frac{dR}{\sqrt{(A(R))}}. \tag{7.2.2}$$

The integral can be evaluated in the general case in terms of elliptic functions, but the result is not easy to understand. Therefore we first examine $R(t)$ qualitatively, and then work out the details of some special cases. It is helpful to consider the acceleration $\ddot{R}(t)$ of the expansion of the universe. From (7.1.4) and (7.1.5), this is

$$\blacktriangleright \quad \ddot{R}(t) = -4\pi\rho_0 G R^3(t_0)/3R^2(t) + (\Lambda/3)\, R(t). \tag{7.2.3}$$

First we examine the *past history* of the models. If the cosmical constant Λ is *zero or negative*, $\ddot{R}$ is negative for all R. Thus the curve of R v. t is convex away from the t axis (figure 52). At some time in the past, therefore, the curve must have intersected the t axis – that is, R must have been zero, and the density ρ infinite. It is natural to call this the moment of the 'origin of the universe', and define the corresponding value of t as zero. Thus the present cosmic time t_0 is the 'age of the universe'. Because $\ddot{R}$ is negative, $\dot{R}$ was greater in the past, and the contraction into the past is faster than it would be if $\dot{R}$ remained constant at its present value. Therefore the

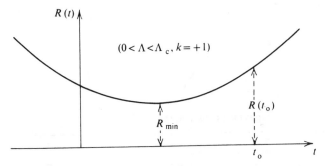

Figure 53. Universe expanding after contraction to finite 'radius'.

universe is *younger* than $1/H_0$, i.e.

▶ $$t_0 < 1/H_0. \qquad (7.2.4)$$

An important restriction on models of this kind is that t_0 must not be too small, otherwise there would be a conflict with astrophysical and geophysical evidence: we would be describing a universe younger than the objects in it!

If Λ is *positive*, $\ddot{R}$ is not always negative, and the possibility arises of $R(t)$ decreasing in the past to a finite minimum R_{min} rather than zero, and increasing into the more distant past. That is, the universe may have contracted to a finite density and then expanded to its present state. At a minimum of $R(t)$ the 'velocity' $\dot{R}(t)$ is zero. Thus, whenever the function $A(R)$ in (7.2.1) has a zero, the corresponding universe did not necessarily explode from an infinitely-dense state. If k is 0 or -1, all the terms in $A(R)$ are positive, so there is no zero of $A(R)$ and the universe must have begun with a 'big bang' (there are a few special cases where this occurred infinitely long ago). The only remaining case is when k is $+1$ (figure 53). Then $\ddot{R}$ can be zero if the single minimum of $A(R)$ is negative. This minimum is easily found to occur at R_c, where

▶ $$R_c^3 = 4\pi G \rho_0 R^3(t_0)/\Lambda, \qquad (7.2.5)$$

and the corresponding value of A is

▶ $$A(R_c) = (4\pi G \rho_0 R^3(t_0))^{\frac{2}{3}} \Lambda^{\frac{1}{3}} - c^2, \qquad (7.2.6)$$

so that a 'big bang' can be avoided only if $\Lambda < \Lambda_c$, where

▶ $$\Lambda_c = c^6/(4\pi G \rho_0 R^3(t_0))^2. \qquad (7.2.7)$$

Therefore, out of all the Friedmann models, based on (7.2.1), the only universes that need not begin with a phase of infinite density are those for which $k = +1, 0 < \Lambda < \Lambda_c$.

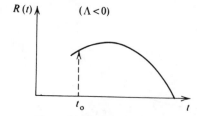

Figure 54. Universe expanding, then collapsing.

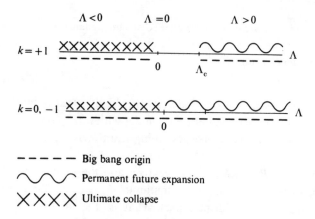

– – – – – Big bang origin

⌢⌣⌢⌣ Permanent future expansion

✕✕✕✕ Ultimate collapse

Figure 55. Beginnings and ends of model universes for different Λ and k.

Similar methods can be employed to examine the *future evolution* of the models If Λ is *negative*, (7.2.3) shows that $\ddot{R}$ is always negative, and the convex curve of $R(t)$ must intersect the t axis at some time in the future as well as in the past. Therefore the present expansion must slow down, stop, and be succeeded by a contracting phase that will end in a gigantic implosion (figure 54). If Λ is *zero*, $\ddot{R}$ is negative for finite R, but zero if R is infinite; therefore it is possible for the expansion to slow down gradually without ever being followed by a contraction. We shall examine the case $\Lambda = 0$ in detail presently, and discover that ultimate collapse is avoided if $k = -1$ or zero, but not if $k = +1$.

If Λ is *positive* $\ddot{R}$ need not be negative, and we must examine $\dot{R}$ using (7.2.1). If k is -1 or zero, $\dot{R}$ can never vanish, so the expansion continues forever. If k is $+1$, it is possible for $\dot{R}$ to vanish if $\Lambda < \Lambda_c$, where Λ_c is given by (7.2.7); then the expansion may be followed by a contraction.

These general properties of model universes can be summed up in figure 55. The unshaded region $0 < \Lambda < \Lambda_c$ for $k = +1$ corresponds to uni-

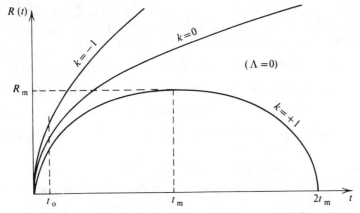

Figure 56. Histories of model universes with zero cosmical constant.

verses that may or may not start or end with a bang, depending on present conditions.

Now we consider in detail various special types of Friedmann model. We examine universes with zero cosmical constant Λ (case A), universes with flat position space ($k = 0$, case B), empty universes ($\rho = 0$, case C) and static universes (case D).

Case A: $\Lambda = 0$. This is the most important case. Since all universes with $\Lambda = 0$ have a 'big bang' origin, we can write the solution (7.2.2) of (7.2.1) as

$$\blacktriangleright \quad t = \int_0^{R(t)} \frac{dR}{\sqrt{(8\pi G\rho_0 R^3(t_0)/3R - kc^2)}}$$

$$= \begin{cases} \dfrac{R_m}{c}\left\{\arcsin\sqrt{\dfrac{R}{R_m}} - \sqrt{\left[\dfrac{R}{R_m}\left(1-\dfrac{R}{R_m}\right)\right]}\right\} & (k = +1) \\[3ex] \dfrac{2R_m}{3c}\left(\dfrac{R}{R_m}\right)^{\frac{3}{2}} & (k = 0) \\[3ex] \dfrac{R_m}{c}\left\{\sqrt{\left[\dfrac{R}{R_m}\left(1+\dfrac{R}{R_m}\right)\right]} - \operatorname{arcsinh}\sqrt{\dfrac{R}{R_m}}\right\} & (k = -1). \end{cases} \quad (7.2.8)$$

The distance R_m is

$$\blacktriangleright \quad R_m \equiv 8\pi G\rho_0 R^3(t_0)/3c^2 = 2q_0 c/H_0|2q_0-1|^{\frac{3}{2}} \quad (k \neq 0)$$
$$= R^3(t_0) H_0^2/c^2 \qquad (k = 0), \qquad (7.2.9)$$

where we have used (7.1.10) and (7.1.11).

The flat ($k = 0$) and open ($k = -1$) universes expand for ever, while the closed ($k = +1$) universe expands to a maximum radius of curvature R_m

at time t_m, and then collapses back to a dense phase at $2t_m$ (figure 56). As we saw in section 7.1, present data suggest a closed universe if Λ is zero. R_m for this case is given by (7.2.9) with $q_0 \sim +1$ as

▶ $\qquad R_m \approx 2c/H_0 = 3.6 \times 10^{10}$ light-years $= 1.3 \times 10^{10}$ pc. $\qquad$ (7.2.10)

The age t_m of the universe in its maximally expanded form is, from (7.2.8),

▶ $\qquad t_m = \pi R_m/2c = 5.7 \times 10^{10}$ years. $\qquad\qquad$ (7.2.11)

Between the birth and death of this universe, a time $2t_m = 11.4 \times 10^{10}$ years elapses.

What is the present age t_0 of these three universes with zero cosmical constant? To find this, we set $R = R(t_0)$ in (7.2.8). If $k = +1$, we use (7.2.10), and obtain

▶ $\qquad t_0 = \dfrac{2q_0}{H_0|2q_0-1|^{\frac{3}{2}}}\left[\text{arc sin }\sqrt{\dfrac{2q_0-1}{2q_0}} - \dfrac{1}{2q_0}\sqrt{(2q_0-1)}\right]$

$\qquad\qquad = 10^{10}$ years $\sim t_m/6,$ $\qquad\qquad\qquad\qquad$ (7.2.12)

where we have used the current 'best' value for q_0, namely 1. Suppose, however, that the true value of q_0 is $\frac{1}{2}$ (which is well within the error bounds), so that the space of the smoothed-out universe is flat. Then, from (7.2.8) and (7.2.9) the present age would be

▶ $\qquad t_0 = \dfrac{2R_m}{3c}\left(\dfrac{R(t_0)}{R_m}\right)^{\frac{3}{2}} = \dfrac{2}{3H_0} = 1.2 \times 10^{10}$ years. $\qquad$ (7.2.13)

Finally, suppose that the present matter density ρ_0 were really equal to the value ρ_{gal} observed in galaxies. Then, from (7.1.10) this would imply that q_0 would be not 1 but $1/40 = 0.025$. The universe would then be *open* ($k = -1$), and its age, from (7.2.8) and (7.2.9), would be

▶ $\qquad t_0 = \dfrac{2q_0}{H_0|2q_0-1|^{\frac{3}{2}}}\left[\dfrac{\sqrt{(|2q_0-1|)}}{2q_0} - \text{arc sinh }\sqrt{\dfrac{|2q_0-1|}{2q_0}}\right] \sim \dfrac{1}{H_0}$

$\qquad\qquad = 1.8 \times 10^{10}$ years. $\qquad\qquad\qquad\qquad$ (7.2.14)

None of these three values of t_0 is so small as to contradict the evidence from geophysics and astrophysics concerning the age of the universe, so that all three models are tenable.

Einstein himself believed that there is a compelling theoretical reason to take $k = +1$: the universe is then closed, and there is no difficulty in understanding how all the matter in it determines the inertial properties of any region of spacetime – in other words, Mach's principle can be seen to apply, clear and unambiguous. If k is 0 or -1, however, the universe is open and in solving Einstein's field equations there are difficulties with

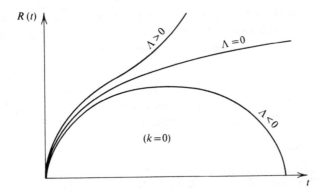

Figure 57. Histories of model universes with flat position space.

'boundary conditions at infinity' which have to be resolved by apparently arbitrary conventions. As we have seen, present evidence in fact does favour a closed universe if $\Lambda = 0$ (which Einstein also believed). With these restrictions, the prediction of general relativity is unique: the universe exploded from a highly condensed state, and will eventually collapse. As Wheeler has said 'No more awe-inspiring prediction has ever been made.'

Case B: $k = 0$. These flat-space universes all began from a condensed state ($R = 0$) so (7.2.1) has the solutions

$$
\blacktriangleright \quad t = \int_0^{R(t)} \frac{\mathrm{d}R}{\sqrt{(8\pi G\rho_0 R^3(t_0)/3R + \Lambda R^2/3)}}
$$

$$
= \begin{cases} \dfrac{2}{\sqrt{(3\Lambda)}} \text{ arc sinh} \left[\left(\dfrac{R(t)}{R(t_0)} \right)^{\frac{3}{2}} \sqrt{\dfrac{\Lambda}{8\pi G\rho_0}} \right] & (\Lambda > 0) \\[3mm] \dfrac{(R(t)/R(t_0))^{\frac{3}{2}}}{\sqrt{(6\pi G\rho_0)}} & (\Lambda = 0) \\[3mm] \dfrac{2}{\sqrt{(3|\Lambda|)}} \text{ arc sin} \left[\left(\dfrac{R(t)}{R(t_0)} \right)^{\frac{3}{2}} \sqrt{\dfrac{|\Lambda|}{8\pi G\rho_0}} \right] & (\Lambda < 0). \end{cases} \quad (7.2.15)
$$

In this case $R(t)$ (figure 57) can be written explicitly as

$$
\blacktriangleright \quad R(t) = \begin{cases} R(t_0) \left(\dfrac{8\pi G\rho_0}{\Lambda} \right)^{\frac{1}{3}} \sinh^{\frac{2}{3}} \left(\tfrac{1}{2}t\sqrt{(3\Lambda)} \right) & (\Lambda > 0) \\[3mm] R(t_0) (6\pi G\rho_0)^{\frac{1}{3}} t^{\frac{2}{3}} & (\Lambda = 0) \\[3mm] R(t_0) \left(\dfrac{8\pi G\rho_0}{|\Lambda|} \right)^{\frac{1}{3}} \sin^{\frac{2}{3}} \left(\tfrac{1}{2}t\sqrt{(3|\Lambda|)} \right) & (\Lambda < 0). \end{cases} \quad (7.2.16)
$$

To apply this to the interpretation of observations, we find Λ from (7.1.7), using (7.1.8) with k zero to eliminate either ρ_0 or q_0, whichever is regarded as more imprecisely known. This gives

▶ $$\Lambda = H_0^2(1-2q_0) = 3H_0^2 - 8\pi\rho_0 G. \tag{7.2.17}$$

If we assume $q_0 = 1$, then $\Lambda = -H_0^2$, so that from (7.2.16) we see that the universe will eventually collapse. The present age is, from (7.2.15) and (7.1.8),

▶ $$t_0 = \frac{2}{H_0\sqrt{(3|1-2q_0|)}} \text{ arc sin } \sqrt{\frac{|1-2q_0|}{2(q_0+1)}}$$

$$\sim \frac{0.6}{H_0} = 1.1 \times 10^{10} \text{ years.} \tag{7.2.18}$$

If we assume on the other hand that the density ρ_0 is more precisely known than q_0, and that ρ_0 equals ρ_{gal}, then (7.1.10) and (7.2.17) give $\Lambda = +2.9H_0^2$ which according to (7.2.16) corresponds to an ever-expanding universe, whose present age is

▶ $$t_0 = \frac{2}{H_0\sqrt{(3\times2.9)}} \text{ arc sinh } \sqrt{\frac{2.9}{7.8/40}} = \frac{1.5}{H_0} = 2.7 \times 10^{10} \text{ years.}$$
$$\tag{7.2.19}$$

Case C: $\rho = 0$. This is the class of *empty* universes, but it might possibly approximate to the actual universe if there is no 'missing mass' and $\rho_0 = \rho_{gal}$, since ρ_{gal} is so small that the galaxies could be considered simply as 'test bodies', their mutual gravitation being negligible. We notice that Λ_c, defined by (7.2.7), is infinite, so that we expect that for $k = +1$ and $\Lambda > 0$ the universe might not begin with a big bang. And indeed from (7.2.2), $R(t)$ is given by

$$t = \int_{c\sqrt{(3/\Lambda)}}^{R(t)} \frac{dR}{\sqrt{(\Lambda R^2/3 - c^2)}} = \left(\sqrt{\frac{3}{\Lambda}}\right) \text{ arc cosh } \left(\frac{R(t)}{c}\sqrt{\frac{\Lambda}{3}}\right),$$

that is,

▶ $$R(t) = c\left(\sqrt{\frac{3}{\Lambda}}\right) \cosh\left(t\sqrt{\frac{\Lambda}{3}}\right), \tag{7.2.20}$$

and this does indeed have the property of expanding not from $R = 0$ but from a finite minimum value $R = c\sqrt{(3/\Lambda)}$.

However, it is more realistic to consider the case $\Lambda < 0$, $k = -1$, which follows from (7.1.7) and (7.1.8) if $\rho_0 = 0$, provided $q_0 > 0$ (as is suggested by observation). Then (7.2.2) gives

▶
$$t = \int_0^{R(t)} \frac{dR}{\sqrt{(c^2 - |\Lambda| R^2/3)}}$$
$$= \left(\sqrt{\frac{3}{|\Lambda|}}\right) \arcsin \left(\frac{R(t)}{c} \sqrt{\frac{|\Lambda|}{3}}\right), \tag{7.2.21}$$

or

▶
$$R(t) = c \left(\sqrt{\frac{3}{\Lambda}}\right) \sin \left(t \sqrt{\frac{|\Lambda|}{3}}\right). \tag{7.2.22}$$

This is another big bang universe that will eventually collapse. To make use of observational data we notice from (7.1.7) and (7.1.8) that

$$R(t_0) \sqrt{|\Lambda|} = \frac{c}{H_0 \sqrt{(q_0 + 1)}} \times \sqrt{(3q_0)} \, H_0 = c \sqrt{\frac{3q_0}{q_0 + 1}},$$

so that the age t_0 can be found from (7.2.21) on setting $R = R(t_0)$. This gives

▶
$$t_0 = \frac{1}{H_0 \sqrt{q_0}} \arcsin \sqrt{\frac{q_0}{q_0 + 1}} = \frac{0.8}{H_0} \sim 1.4 \times 10^{10} \text{ years.} \tag{7.2.23}$$

Case D: static models. This has historical interest only, because the universe is not static but expanding. However, this was not known in 1916, and it was natural for Einstein to seek a static solution of his equations for the first model universe based on general relativity; it was to obtain such a solution that he introduced the cosmical constant Λ. In a static universe, R and ρ are constants, so $\dot{R}$ and $\ddot{R}$ are zero. Then (7.2.3) gives

▶
$$\Lambda = 4\pi\rho_0 G, \tag{7.2.24}$$

and then (7.2.1) gives

$$k/R^2 = 4\pi\rho_0 G/c^2,$$

i.e.

▶
$$k = +1, \ R = c/\sqrt{(4\pi\rho_0 G)}, \tag{4.2.25}$$

where R is the constant value of the scale factor. (These results also follow from (7.1.7) and (7.1.8) if it is realized that Hubble's constant vanishes in a static universe.) If we use the value ρ_{gal} for ρ_0, we obtain, for this model:

▶
$$\Lambda = +(6.1 \times 10^{10} \text{ years})^{-2},$$
$$R = 6.1 \times 10^{10} \text{ light-years} = 1.8 \times 10^{10} \text{ pc.} \tag{7.2.26}$$

Our tour through the various relativistic model universes has not been complete; we have not studied in detail those Friedmann models in which

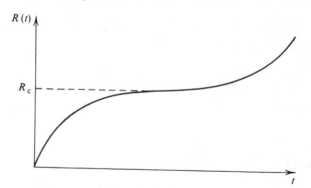

Figure 58. History of Eddington–Lemaître, or 'dachshund', universe.

all three terms in (7.2.1) are non-zero. An interesting model of this type (associated with Eddington and Lemaître) is the 'hesitation' or 'dachshund' universe (figure 58) where Λ is slightly greater than Λ_c (equation (7.2.7)) and $k = +1$; then the curve $R(t)$ has a point of inflexion at R_c (see equations (7.2.3) and (7.2.5)), at which $\dot{R}$ is small. Therefore, on this model there is a long time during which $R(t)$ is very close to R_c; radiation emitted during this time would be received by us with almost the same red shift (equation (6.2.1)), and it has been suggested that this might explain the clustering of QSO red shifts near $z = 2.2$ (figure 51) (this explanation is now thought to be inadequate). Another feature of this model universe is that its age can greatly exceed H_0^{-1}; before 1950, underestimation of cosmic distances meant that H_0^{-1} seemed less than the known ages of the elements and stars and the Eddington–Lemaître model was often invoked as a way of preventing the universe being younger than its contents.

One feature of all universe models except some for $0 < \Lambda < \Lambda_c$ is an explosion from an initial highly condensed state – possibly even a singularity – with $R = 0$. The physical nature of this early dense state will form the subject of chapter 8. Now we shall indicate a possible answer to the question: 'what happened before the big bang?' The explosion at $t = 0$ might have been preceded by a previous collapse, and we can even envisage a whole series of cycles of expansion and collapse (figure 59) in the case of those universes for which $R(t)$ can have two zeros (e.g. $\Lambda = 0$, $k = +1$); each cycle lasts for a time $2t_m$. These are the *oscillating universe* models. At present such models represent an extreme of cosmological speculation. To what extent are the cycles identical? How similar to our own will be the universe at time $t_0 + 2t_m$? What is the physical nature of the 'bounces' at times 0, $\pm 2t_m$, $\pm 4t_m$, ...? Do these bounces conserve Λ and k? Are the laws of physics the same before and after each bounce?

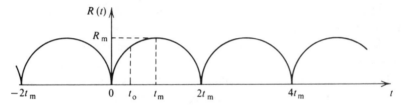

Figure 59. Oscillating model universe.

It has even been suggested that *time runs backwards* in the contracting phases of the universe. This is not as absurd as it first appears. The 'arrow of time' – that is, the distinction between the past and the future – appears nowhere in the fundamental physical laws of mechanics (quantum, relativistic or classical) or electromagnetism. These laws are time-symmetrical, that is, they are invariant under the transformation $t \to -t$. Asymmetry first appears in the second law of thermodynamics, which states that the entropy of an isolated system which is not in equilibrium will always increase; thus the direction of increasing entropy defines a 'thermodynamic arrow of time'. The relaxation to equilibrium is always accompanied by a release of particles or *radiation*, which expands outwards and is lost in the depths of the universe. The time-reversed solution of the equations of electrodynamics, corresponding to a wave contracting on to a body, is almost never encountered; thus, the direction in which waves spread out defines an 'electrodynamic arrow of time'. By arguments like those used in discussing the darkness of the night sky, it is claimed that the *expansion of the universe* (which defines a 'cosmological arrow of time') enables radiation to be 'lost'. This would imply that during a contracting phase light would be absorbed by stars and emitted by eyes, and that entropy would decrease and time run backwards. Quite apart from the logical difficulties of this argument, it is difficult to see by what physical mechanism the sign of the rate of change of entropy could suddenly reverse in the rarefied maximally expanded universe at time t_m. The present position is that there are no well-established relations between the three 'arrows of time' described above.

7.3 The steady state theory

This is founded on the *perfect cosmological principle*, which states that the smoothed-out universe is unchanging as well as spatially homogeneous. In Bondi's words: 'Geography doesn't matter, and history doesn't matter either.' Of all the uniform cosmologies based on the Robertson–Walker separation formula (6.1.3), the steady state model is the simplest, because it demands that all observers will measure the same constant values for averaged quantities such as the expansion rate, deceleration parameter, spatial curvature, matter density and galactic number density. The main empirical motivation for introducing the model was that it describes an infinitely old universe, and thus evades the problem posed by early observations which seemed to imply that H_0^{-1} was smaller than the ages of the elements and stars (cf. the discussion of the Eddington–Lemaître universe – figure 58). It is a remarkable fact, first realised by Bondi and Gold in 1948, that the perfect cosmological principle is enough to determine the form of the cosmic scale factor $R(t)$: the expansion of the universe, described by Hubble's constant, must always occur at the same rate H_0 as we measure today. Thus (6.1.6) gives

$$H_0 = H(t) = \dot{R}(t)/R(t),$$

or

▶ $$R(t) = A \exp (H_0 t), \tag{7.3.1}$$

where A is a constant. From (6.1.1), the spatial curvature is then

$$K(t) = k/R^2(t) = (k/A^2) \exp (-2H_0 t).$$

But $K(t)$ must be constant, and this can happen only if

▶ $$k = 0. \tag{7.3.2}$$

Thus the position space of the steady state model is *flat*. This does not mean that *spacetime* is flat, and indeed (7.1.1) shows that its curvature $\mathcal{K}$ is given by

▶ $$\mathcal{K} = -\ddot{R}(t)/R(t) = -H_0^2. \tag{7.3.3}$$

The deceleration parameter q_0 can easily be calculated from the definition (6.2.3), and we have

▶ $$q_0 = -\ddot{R}/RH_0^2 = -1. \tag{7.3.4}$$

This value of q_0 brings the steady state theory into a conflict with observations of apparent magnitudes and red shifts of galaxies. As discussed in section 6.3, these observations are best fitted with $q_0 \sim 1$ (see also figure 6). This conclusion is based on equations (6.3.6) and (6.3.7), which are in turn derived assuming that the average absolute luminosity L of a galaxy is independent of time. In evolutionary models, such as those described in section 7.2, this assumption is highly dubious, and indeed the analysis of number counts in section 6.5 shows that it is necessary to correct for the evolution of radio sources. However, the steady state model cannot evade a possible disagreement with experiment by appealing to possible changes in L, because that would contradict the basic postulates of the model;

individual sources evolve and die, of course, but new ones are constantly condensing, and the average value of L is constant.

Now we consider the *mass density* ρ. If matter is conserved, this density must change with time according to (7.1.5); as $R(t)$ changes, so must ρ. But $R(t)$ *does* change on the steady state model, according to (7.3.1). However, the perfect cosmological principle requires that ρ must be constant. Therefore to keep the steady state model we must give up matter conservation, and postulate that *matter is created* at such a rate as to keep the density constant in an expanding universe. This was the approach followed by Hoyle in 1948; he added a 'creation' term to the dynamical equation (7.1.4) of general relativity, and solved for $R(t)$, thus obtaining an alternative derivation of the result (7.3.1).

As a result of the continuous creation of matter, the number density n of sources will remain constant. This means that when using the steady state theory to interpret counts of the number $N(> l)$ of sources with apparent luminosity exceeding l, we cannot use the formula (6.5.4), since that was derived on the assumption that $n(t)$ obeyed the conservation law (6.4.2). To obtain $N(> l)$ for a steady state universe, we start with the general formula (6.5.1) for $N(> t_e)$ – that is, for the number of sources now visible that emitted their radiation after cosmic time t_e. Setting $n(t) = n = $ constant and using (7.3.1) for $R(t)$, we obtain

$$N(> t_e) = 4\pi n c^3 \int_{t_e}^{t_o} dt (A \exp(H_o t))^2 \left[\int_t^{t_o} \frac{dt'}{A} \exp(-H_o t') \right]^2$$

$$= \frac{4\pi n c^3}{H_o^2} \int_{t_e - t_o}^0 d\tau (\exp(+H_o \tau) - 1)^2.$$

Expansion for small $t_e - t_o$ gives

$$N(> t_e) = \frac{4\pi n c^3}{3} (t_o - t_e)^3 \left[1 - \frac{3H_o}{4} (t_o - t_e) + ... \right].$$

Now we use the red shift expansion (6.2.5) with $q_0 = -1$, and we obtain, for the number of sources with red shifts less than z:

$$\blacktriangleright \qquad N(< z) = \frac{4\pi n c^3 z^3}{3 H_o^3} \left(1 - \frac{9z}{4} + ... \right). \qquad (7.3.5)$$

This is the steady state analogue of (6.5.3). Finally, we obtain $N(> l)$ by expressing z in terms of l using (6.3.6), with $q_0 = -1$; this gives

$$\blacktriangleright \qquad N(> l) = \frac{4\pi n}{3} \left(\frac{L}{4\pi l} \right)^{\frac{3}{2}} \left[1 - \frac{21 H_o}{4c} \left(\frac{L}{4\pi l} \right)^{\frac{1}{2}} + ... \right]. \qquad (7.3.6)$$

This formula is the steady state analogue of (6.5.4). As with (6.5.4), the correction term is negative, so there should be fewer faint sources than the elementary $l^{-\frac{3}{2}}$ law predicts. As we saw in section 6.5, observation indicates just the opposite. In time-dependent models this is interpreted as evidence that the average luminosities of sources change as they evolve. The steady state theory forbids this conclusion, since the *average* luminosity L must be kept constant by the creation of matter that condenses into new sources to replace those that die. Much more than the measurements on q_0, this evidence from number counts has convinced most cosmologists that the steady state theory, although elegant and simple, is no longer tenable. In chapter 8 we shall see how the microwave radiation background strongly suggests an evolving universe, and therefore constitutes further evidence against the steady state theory.

7.4   Cosmologies in which the strength of gravity varies

In 1938 Dirac devised a cosmology based on a remarkable numerical coincidence among the fundamental quantities of physics. On a *cosmological* scale the important quantities are Hubble's constant H_0 and the gravitational constant G; their ratio is G/H_0. On a *microscopic* scale the important quantities are Planck's constant h and the charge e and mass m_p of the proton; from these, if we include the 'unit-conversion' factor ϵ_0 (permittivity of free space), we can form the combination

$$h^3 \epsilon_0 / m_p^3 e^2,$$

which has the same dimensions as G/H_0 (mass^{-1} length^3 time^{-1}). The dimensionless ratio of these microscopic and macroscopic quantities is

▶ $$\frac{G}{H_0} \div \frac{h^3 \epsilon_0}{m_p^3 e^2} = 1.3. \qquad (7.4.1)$$

Considering the enormous range of numerical values of the various quantities, this is astonishingly close to unity. In place of e, we could introduce the speed of light via the relation

$$\frac{e^2}{4\pi \epsilon_0 \hbar c} \approx 137,$$

or, we could argue that $4\pi\epsilon_0$ should appear in (7.4.1) rather than just ϵ_0. But whatever we do, the ratio $G/H_0 \div$ (combination of microscopic quantities) remains of order unity. (There is another dimensionless combination of microscopic quantities and G, not involving H_0, namely

$$\frac{G m_p^2 \epsilon_0}{e^2}$$

whose value – about 10^{-35} – is very different from unity.) Dirac's argument was that the relation (7.4.1) might not be an accident, but might instead represent a deep and as yet unexplained connection between cosmic and microscopic physics.

However, Hubble's 'constant' changes with time for most model universes, so that if (7.4.1) is to represent a law of nature, one or more of the fundamental 'constants' G, h, m_p and e must change with time. This conclusion is not inevitable: for oscillating universes it could be claimed that (7.4.1) should contain not $1/H_0$ but the time $2t_m$ of an explosion-implosion cycle; the equation would then relate true constants of nature. Nevertheless, it is interesting to pursue the possibility that (7.4.1) is correct as it stands, so that the 'constants' cannot be constant. The simplest procedure is to allow G to vary and keep h, m_p and e fixed. The reason is that any variation in the microscopic constants would imply that atomic and nuclear physics, and hence the structure of materials and their chemistry, etc., were different in the past, and there is no evidence (e.g. from the spectra of distant galaxies) that this was the case. Therefore we assume that the gravitational 'constant' $G(t)$ is proportional to Hubble's 'constant' $H(t)$. Now $H(t)$ is roughly equal to the reciprocal of t, the age of the universe (assuming time to be reckoned from the moment of birth). Therefore Dirac's theory predicts

▶ $\qquad G(t) \propto 1/t.$ (7.4.2)

This result can be obtained in another way. If there is 'missing mass' between the galaxies with a density of about $40\rho_{gal}$, then (7.1.10) shows that the present density ρ_0 is related to G and H_0 by

▶ $\qquad \rho_0 G/H_0^2 \sim \frac{1}{4}.$ (7.4.3)

This is another numerical coincidence, this time between purely cosmological numbers, and Dirac assumed that it too represents a fundamental law and therefore still holds as the universe expands and ρ decreases according to (7.1.5). Therefore

$$\frac{\rho(t)\,G(t)}{H^2(t)} \propto \frac{1}{R^3(t)} \times G(t) \times \frac{R^2(t)}{\dot{R}^2(t)} = \frac{G(t)}{R(t)\,\dot{R}^2(t)} = \text{const.}$$

However, the original relation (7.4.1) gives

$$\frac{G(t)}{H(t)} = \frac{G(t)\,R(t)}{\dot{R}(t)} = \text{const.}$$

$G(t)$ can be eliminated, and we obtain

$$R^2(t)\,\dot{R}(t) = \text{const.}$$

so that

$$R(t) \propto t^{\frac{1}{3}}.$$

Differentiating once, we obtain

▶ $$H(t) = 1/3t,$$ (7.4.4)

from which (7.4.2) follows.

Now we compare Dirac's cosmology with observation. From (6.2.3) we obtain for q_0 the value $+2$. This is just possible, according to the current estimate of 1 ± 1. From (7.4.4) we obtain for the present age of the universe the value

▶ $$t_0 = 1/3H_0 = 6 \times 10^9 \text{ years.}$$ (7.4.5)

However, this would imply that the universe is younger than the elements, whose age has been established as about 10^{10} years by radioactive dating techniques (which depend on nuclear processes and so are unaffected by changes in G). Thus Dirac's universe is not old enough.

For further evidence, we examine some consequences of G being larger in the past than it is now. Firstly, the stars would have had to be brighter, basically because they would need to radiate faster in order for radiation pressure to prevent gravitational collapse. In fact, approximate theories of opacity and energy release indicate that the absolute luminosity $L(t)$ at a time t would have been greater by a factor $(G(t)/G(t_0))^8$ than if G had remained constant. In addition, the Earth's distance from the Sun, $r_\oplus(t)$, would have been smaller by a factor $G(t)/G(t_0)$. Therefore the apparent luminosity $l(t)$ of the Sun would have been greater than it is now by a factor

$$\frac{l(t)}{l(t_0)} = \frac{L(t)/4\pi r_\oplus^2(t)}{L(t_0)/4\pi r_\oplus^2(t_0)} = \left(\frac{G(t)}{G(t_0)}\right)^{10}.$$

This means that the temperature $T(t)$ of the Earth's surface would have been greater than it is now, because since the Earth is in equilibrium with the radiation it receives from the Sun we can use Stefan's law, to obtain

Total emissive power $\propto l \propto T^4$.

Thus we have

▶ $$\frac{T(t)}{T(t_0)} = \left(\frac{l(t)}{l(t_0)}\right)^{\frac{1}{4}} = \left(\frac{G(t)}{G(t_0)}\right)^{2.5} = \left(\frac{t_0}{t}\right)^{2.5},$$ (7.4.6)

since $G(t) \propto t^{-1}$. Now we assume, consistently with (7.4.5), that the age of the Earth is 5×10^9 years. This means that a thousand million years ago the Earth was about four-fifths of its present age, so, if we take the present temperature as 300 K, we have

$$T(4 \times 10^9 \text{ years}) = \left(\frac{5}{4}\right)^{2.5} \times T(t_0) = 1.75 T(t_0) \approx 530 \text{ K}.$$

This is far too hot for life to have developed. Even if we take the present temperature as 273 K we find that any surface water would have boiled as recently as six hundred million years ago. We conclude that it is difficult to reconcile Dirac's cosmology with data from geophysics and palaeontology

The idea that the strength of gravity varies is the basis of a fully relativistic theory developed principally by Brans and Dicke. The theory involves spacetime and geodesics, and generates uniform model universes with the Robertson–Walker separation formula (6.1.3). However, the dynamical equations connecting matter with the curvature of spacetime are no longer the 'simplest possible', as in Einstein's theory, but involve an extra constant of nature, ω, whose effect is to make G vary with time. The introduction of ω was not an *ad hoc* complication of general relativity; it was hoped that the Brans–Dicke theory would satisfactorily incorporate Mach's principle into relativity (the extent to which Einstein's theory does this is still uncertain). However, precise experiments such as Shapiro's measurements of radar time delay (section 5.5) have narrowed down the range of possible values of ω to the point where the Brans–Dicke theory is observationally indistinguishable from Einstein's. Therefore any variation of G must be extremely small, so that the new theory cannot explain the numerical coincidence (7.4.1). Moreover, any geophysical effects would be virtually impossible to detect.

8 *In the beginning*

8.1 Cosmic black-body radiation

The universe contains radiation in the form of photons moving in all directions with all frequencies. Cosmic expansion causes the energy density $\epsilon_r(t)$ of this radiation to decrease, as follows: the number density of photons decreases as $R^{-3}(t)$ (because volumes expand as $R^3(t)$), and the energy of individual photons decreases as $R^{-1}(t)$ (because of the red shift in frequency). Therefore $\epsilon_r(t)$ decreases as $R^{-4}(t)$, that is,

$$\blacktriangleright \qquad \epsilon_r(t) = \frac{\epsilon_r(t_0)\, R^4(t_0)}{R^4(t)}. \qquad\qquad (8.1.1)$$

The equivalent *mass* density $\rho_r(t)$ is

$$\blacktriangleright \qquad \rho_r(t) = \frac{\epsilon_r(t)}{c^2}, \qquad\qquad (8.1.2)$$

and therefore also decreases as $R^{-4}(t)$.

According to (7.1.5), the mass density $\rho(t)$ of the *matter* in the universe decreases as $R^{-3}(t)$. Therefore as we go backwards in time the radiation density increases faster than the matter density, so that, however small ρ_r is now, there must have been a time t_E when the densities of matter and radiation were equal. At this time, we have

$$\rho(t_E) = \frac{\rho_0\, R^3(t_0)}{R^3(t_E)} = \rho_r(t_E) = \frac{\rho_r(t_0)\, R^4(t_0)}{R^4(t_E)},$$

that is,

$$\blacktriangleright \qquad \frac{R(t_0)}{R(t_E)} = \frac{\rho_0}{\rho_r(t_0)}. \qquad\qquad (8.1.3)$$

At times earlier than t_E, ρ_r was greater than ρ, and the period $0 < t < t_E$ is therefore called the *radiation-dominated* era of the history of the universe. The contents of the radiation-dominated universe are referred to as the 'primaeval fireball', or as 'ylem' (from the Greek 'hyle' meaning 'that on which form has yet to be imposed').

In the radiation-dominated era there is, as we shall see, good reason to believe that the radiation and matter were in thermal equilibrium at the (same) temperature $T_r(t)$. Therefore the radiation must have had a *black-body distribution*. As the universe expanded, the mean photon energy decreased as $R^{-1}(t)$, so that the temperature, which is proportional to this mean energy, should also have decreased as $R^{-1}(t)$. At the time t_E the universe entered the *matter-dominated* era. As we shall discuss later, it so happens that at about the same time the radiation and matter ceased to be in thermal equilibrium; thus $T_r(t)$ need no longer be equal to the matter temperature $T_m(t)$. A detailed theoretical study has been made of the transition between the radiation-dominated and matter-dominated eras. This strongly suggests that no significant distortion of the black-body distribution occurs during the decoupling of matter and radiation near the time t_E. If the radiation preserves its black-body character *after* t_E (when it is no longer coupled to matter), then $T_r(t)$ must continue to be proportional to $R^{-1}(t)$, so that, if T_{ro} is the present radiation temperature,

$$\blacktriangleright \qquad T_r(t) = T_{ro}\, \frac{R(t_0)}{R(t)}. \qquad (8.1.4)$$

Now we prove that the radiation should indeed preserve its black-body character. The proof is based on the conservation of photon number: this number could only be significantly altered by interactions involving matter, but the number of elementary particles of matter (nucleons, electrons etc.) is easily calculated to be far smaller ($\sim 10^{-10}$) than the number of photons – even though the matter now dominates energetically and gravitationally. According to Planck's law, the number $dN(t)$ of photons with frequencies between ν and $\nu+d\nu$ in a volume $V(t)$ of space at cosmic time t is

$$\blacktriangleright \qquad dN(t) = \frac{8\pi\nu^2 V(t)\, d\nu}{c^3\,(\exp(h\nu/kT_r(t))-1)}. \qquad (8.1.5)$$

As time proceeds, the number of photons in the volume remains the same, because of conservation, which implies that a comoving observer would see as many photons crossing the imaginary boundary into the volume as leaving it. At a new time t', then, the original group of photons has been red-shifted to a frequency

$$\nu' = \frac{\nu R(t)}{R(t')}, \quad d\nu' = d\nu\, \frac{R(t)}{R(t')},$$

and the volume has expanded to

$$V(t') = V(t)\frac{R^3(t')}{R^3(t)}.$$

Therefore for this group of photons we have

$$dN(t') = dN(t) = \frac{\dfrac{8\pi}{c^3}\left(\dfrac{\nu' R(t')}{R(t)}\right)^2 V(t')\dfrac{R^3(t)}{R^3(t')}\,d\nu'\,\dfrac{R(t')}{R(t)}}{\exp(h\nu' R(t')/R(t)\, T_r(t)\, k) - 1}$$

$$= \frac{8\pi V(t')\,\nu'^2\,d\nu'}{c^3(\exp(h\nu'/kT_r(t')) - 1)},$$

where the new temperature is

$$T_r(t') = T_r(t)\, R(t)/R(t').$$

Thus the radiation keeps its black-body distribution even when it is no longer in equilibrium with the matter; it expands adiabatically with the universe, and cools in accordance with equation (8.1.4). The energy density $\epsilon_r(t)$ is obtained from (8.1.5) by setting $V(t) = 1$ and integrating:

$$\epsilon_r(t) = \int dN h\nu = \frac{8\pi h}{c^3}\int_0^\infty d\nu\, \nu^3/(\exp(h\nu/kT_r(t)) - 1)).$$

The integral is standard, and the final formula for $\epsilon_r(t)$ is

$$\blacktriangleright \qquad \epsilon_r(t) = 8\pi^5 k^4 T_r^4(t)/15c^3h^3 \equiv aT_r^4(t), \qquad (8.1.6)$$

where k is Boltzmann's constant and $a\ (= 7.5\times 10^{-16}\ \mathrm{J\ m^{-3}\ K^{-4}})$ is the Stefan–Boltzmann constant. This is perfectly consistent with (8.1.1), since according to (8.1.4) T_r is proportional to R^{-1}.

This theoretical discussion suggests that the universe should contain black-body radiation, and the present evidence is that it does. The first observation of the cosmic microwave background, by Penzias and Wilson in 1965, was the most important cosmological discovery since Hubble established the expansion of the universe. The data obtained so far are summarized in figure 60. It is difficult to measure the high-frequency part of the curve, because the frequencies of the corresponding radiation lie outside the range for which the atmosphere is a transparent window. Therefore an intensive series of balloon and rocket experiments is currently investigating the frequency distribution beyond the maximum of the black-body curve. The present data, however, do fit very closely to a black-body distribution with temperature

$$\blacktriangleright \qquad T_{r0} = 2.7\ \mathrm{K}. \qquad (8.1.7)$$

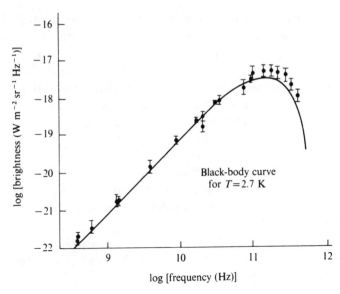

Figure 60. Spectrum of cosmic microwave background radiation.

This means that most of the radiation has wavelengths of the order of millimetres – i.e. the radiation lies in the microwave region of the spectrum. From (8.1.6) we obtain, for the present value of the energy density,

$$\blacktriangleright \qquad \epsilon_r(t_0) = aT_{ro}^4 = 4 \times 10^{-14} \text{ J m}^{-3}. \qquad (8.1.8)$$

The equivalent mass density $\rho_r(t_0)$ is, from (8.1.2),

$$\blacktriangleright \qquad \rho_r(t_0) = 4.5 \times 10^{-31} \text{ kg m}^{-3}, \qquad (8.1.9)$$

so that the radiation density is only about a thousandth of the currently accepted value (2.1.2) for the matter density ρ_0, and the universe is indeed matter-dominated at the present time. Of course the universe contains not just microwaves but all other kinds of radiation; however, this is not distributed with a black-body spectrum, and not nearly so intense as the microwave background (the total non-black-body energy density has been estimated as less than $\epsilon_r(t_0)/100$). There have been attempts to save the steady state theory by trying to explain the microwave radiation as coming from a great number of discrete distant sources, but all these explanations make *ad hoc* assumptions about the spectral characteristics of these sources. To sum up, most cosmologists believe that the microwave radiation does indeed have a black-body character, and so is an adiabatically-cooled, redshifted pale remnant of the primaeval fireball.

The discovery that the background radiation exists opened up the fascinating possibility of measuring the velocity of the Earth relative to the cosmic fluid, that is, relative to the average motion of the local matter of the universe. To a comoving observer the radiation would appear isotropic by definition. Therefore another observer in motion relative to the first would see the radiation coming from his forward direction blue-shifted, and the radiation from his·backward direction red-shifted. In other words, the black-body radiation temperature would appear to be direction-dependent. Such anisotropy has indeed been found, and it is concluded that the Earth is moving relative to the cosmic fluid with a speed of about a few hundred kilometres per second. We emphasise that this motion is not 'absolute' or 'relative to the ether' but relative to the perfectly well defined frame of reference with respect to which cosmic matter is locally at rest, on the average.

Now we study the transition between the radiation- and matter-dominated eras. From (8.1.3), if we take the values (8.1.9) and (2.1.2) for $\rho_r(t_0)$ and ρ_0, we get

▶ $\qquad R(t_0)/R(t_E) \approx 700.$ $\hfill$ (8.1.10)

(We have used (2.1.2), which is the lowest estimate of ρ_0; if 'missing mass' exists, then $R(t_0)/R(t_E)$ will be greater than 700.) The transition temperature $T_r(t_E)$ is, from (8.1.4),

▶ $\qquad T_r(t_E) \approx 1900 \text{ K}.$ $\hfill$ (8.1.11)

(Again, if there is 'missing mass' $T_r(t_E)$ will be greater than this.) Now, this value of 1900 K happens to be of the same order as 4000 K, above which it has been calculated that a dilute gas of hydrogen would be almost completely ionised. Thus the transition period $t \approx t_E$, between the radiation- and matter-dominated eras, is also the *recombination era* during which the plasma of free electrons and protons condenses into neutral hydrogen. However, the plasma is far more opaque to radiation than is the neutral hydrogen (which is virtually transparent). Therefore, before t_E matter interacted far more strongly with radiation than it does today, and this is the basis of our assertion that radiation was once in equilibrium with matter and so acquired a black-body distribution.

In the radiation-dominated era, cosmic dynamics based on general relativity is particularly simple. The governing equation is (7.1.6); for small t, $R(t)$ is also small and $\rho(t)$ is large, and the terms involving the curvature index and the cosmical constant are negligible. Thus we have, for the early history of all exploding models,

$\dot{R}^2(t) = 8\pi G\rho(t) R^2(t)/3.$

This is a relativistic equation, so that $\rho(t)$ must be interpreted as the total mass density of matter *and radiation*; for $t \ll t_E$, ρ_r dominates. Thus we can use (8.1.2) and (8.1.6) to express ρ_r in terms of T_r, and use (8.1.4) to express R and hence $\dot{R}$ in terms of T_r; we obtain the equation

$$\dot{T}_r^2 = (8\pi Ga/3c^2)\, T_r^6,$$

whose solution is

► $\qquad T_r(t) = (3c^2/32\pi Ga)^{\frac{1}{4}}/\sqrt{t}.$ (8.1.12)

For the radiation density, we obtain

► $\qquad \rho_r(t) = 3/32\pi Gt^2.$ (8.1.13)

It should be noted that these formulae for the early thermal history of the universe contain no adjustable constants, and apply irrespective of the values of Λ and k. For the cosmic scale factor, (8.1.4) and (8.1.12) give

► $\qquad R(t) = R(t_0)\, T_{r0}(32\pi Ga/3c^2)^{\frac{1}{4}} \sqrt{t}.$ (8.1.14)

It is not strictly correct to apply these formulae to the transition era $t \approx t_E$, because then the matter content of the universe is no longer negligible, and in addition the terms in (7.1.6) involving Λ and k might be significant. However, if we are interested only in rough estimates, we can solve (8.1.12) and (8.1.11) for t_E, and we obtain

► $\qquad t_E = \left(\sqrt{\dfrac{3c^2}{32\pi Ga}} \right) \dfrac{1}{T_r^2(t_E)} \approx 6 \times 10^{13}\,\text{s} \approx 2 \times 10^6\,\text{years}.$ (8.1.15)

If we take 10^{10} years as a rough value for the age of the universe, it is clear that the universe was radiation-dominated for only one ten-thousandth part of its history. In fact this is probably an overestimate: if 'missing mass' is discovered, the value $T_r(t_E)$ will be greater, and t_E smaller.

8.2 Condensation of galaxies

So far we have explored the 'uniform' models of the universe, that is, those models in which cosmic matter and radiation are homogeneously distributed. However, on scales less than tens of megaparsecs matter is very inhomogeneously distributed, in the form of galaxies and stars. In this section we consider the question 'How did the galaxies originate?' Most theories assume that the galaxies condensed as a result of self-gravitation, but the problem has been to find a plausible description of the history of this condensation, and especially to understand why galaxies have the range of masses measured today ($10^6\, M_\odot$ to $\sim 10^{12} M_\odot$) and particularly why there is a rather sharp upper limit. It is only recently that the outlines of a satisfactory theory have begun to emerge.

Galaxies probably condensed from primordial density inhomogeneities. How did these inhomogeneities arise? Was an initially smooth ylem 'stirred by the finger of God', were the inhomogeneities part of the initial conditions of the big bang, or what? If the elementary particles of the matter in the universe started out with a uniform distribution, then purely statistical fluctuations would occur on all scales, from which matter would eventually condense. However, the development of highly improbable fluctuations of galactic dimensions ($\sim 10^{68}$ particles) would be extremely slow, and it has been calculated that the universe is much too young for this to be a plausible model. There is also an objection in principle to any theory postulating initially homogeneous matter: it implies that regions of the universe which could never have been causally connected would, unreasonably, have the same density. To see this, consider the comoving coordinate σ_{oh} of the object horizon introduced in section 6.2 (equation (6.2.6)); for short times, we see from (8.1.14) that

$$\sigma_{oh} \propto t^{\frac{1}{2}}.$$

Thus the size of the observable universe, and hence the range of causal influence, increases from zero. (The mass M_{oh} of matter inside the object horizon at cosmic time t is easily calculated to be

$$M_{oh} = (c^3 t/G) \sqrt{(t/t_E)},$$

and corresponds to a large galactic mass when $t \sim 1$ year.)

For these reasons, considerable effort is now being directed towards studying models with 'total initial chaos', that is, models with initial density inhomogeneities on all scales, distributed according to some as-yet-unknown law. Some limits on the strengths of the initial inhomogeneities (especially on larger scales) are set by the observed degree of isotropy of the microwave radiation background: large density differences might have been associated with large fireball temperature differences, which would still be observable today in fossilised form as anisotropies in the microwave radiation distribution.

We shall assume, then, that at some time in the early universe (perhaps $t = 0$) there existed weak density inhomogeneities on all scales. What physical processes govern their development? Basically three: *Gravitation* is a source of *instability*, causing an inhomogeneity, denser than the background, to tend to contract and become denser still (figure 61(*a*)). *Cosmic fluid pressure* is a source of *oscillations*, causing an inhomogeneity to radiate away as acoustic waves (figure 61(*b*)), analogous to the waves arising when the surface of a pond is disturbed. *Photon viscosity* is a source of *damping*, causing an inhomogeneity to decay away; it arises from the

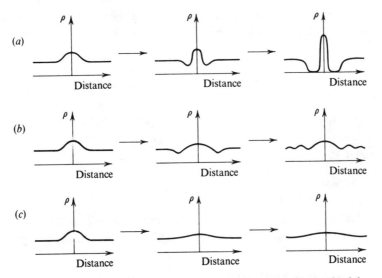

Figure 61. Three effects contributing to development of a density inhomo-geneity: (*a*) gravitational instability; (*b*) cosmic fluid pressure oscillations; (*c*) photon viscosity causing damping.

drag force on the matter resulting from collisions with photons (figure 61(*c*)).

We consider first the competition between gravitational instability and pressure oscillations for an inhomogeneity of mass M (this refers to mass in the form of matter, not radiation). The theory was first worked out by Jeans in 1902. If M exceeds a certain critical mass M_J (the 'Jeans mass'), the inhomogeneity will be unstable and contract, because the elastic restoring forces arising from the fluid pressure will be weaker than the self-gravitation of the mass. If M is less than M_J the elastic forces will dominate and the density will oscillate. Obviously M_J will depend on the condition of the fluid, that is, on its density, temperature and whether it is matter or radiation dominated. We shall now estimate M_J as a function of time.

An inhomogeneity with 'radius' r will, if it oscillates, resemble a sound wave with wavelength r. The characteristic 'restoring time' is the wave period r/v, where v is the speed of the sound in the fluid. If r/v exceeds the characteristic time for the disturbance to contract under its own gravity, the disturbance is unstable and contracts; otherwise it oscillates. The characteristic contraction time could depend only on G, r and the *total* density (matter + radiation) ρ_{tot}; in fact the only function with the dimen-sion 'time' that can be constructed from these quantities is const. $\times (G\rho_{tot})^{-\frac{1}{2}}$

which happens not to involve r. We shall take the value of the 'constant' as unity (a full analysis gives $\pi^{\frac{1}{2}}$). Thus the condition for contraction is

$$r/v > 1/\sqrt{(G\rho_{\text{tot}})},$$

so that the Jeans mass is

▶ $$M_{\text{J}} = 4\pi\rho v^3/3(G\rho_{\text{tot}})^{\frac{3}{2}}, \qquad (8.2.1)$$

where ρ is the density of the matter alone. To calculate M_{J} we need v; if p is the gas pressure (which depends on ρ_{tot} and T via the equation of state) then

▶ $$v = \sqrt{\frac{\mathrm{d}p}{\mathrm{d}\rho_{\text{tot}}}}. \qquad (8.2.2)$$

We deal separately with radiation- and matter-dominated eras.

In the *radiation-dominated era* $(t < t_{\text{E}})$, the matter density is negligible, and we have

$$\rho_{\text{tot}} \approx \rho_{\text{r}} = aT_{\text{r}}^4/c^2,$$

where we have used (8.1.2) and (8.1.6). The equation of state is

$$p = \rho_{\text{r}}c^2/3;$$

this can be justified as follows: one-third of the radiation is moving in any particular direction, with speed c, so that the pressure, which is the momentum transferred in unit time across unit area, is mass crossing $\times c$, i.e. $(\frac{1}{3}\rho_{\text{r}} \times c) \times c$, which is what we wanted to show. Thus from (8.2.2) we have, for the speed of sound in the photon gas,

$$v = \sqrt{\frac{\mathrm{d}p}{\mathrm{d}\rho}} = \frac{c}{\sqrt{3}},$$

and the Jeans mass (8.2.1) is

$$M_{\text{J}} = \frac{4\pi\rho c^3}{3(3GaT_{\text{r}}^4/c^2)^{\frac{3}{2}}} \approx \frac{\rho c^6}{T_{\text{r}}^6(Ga)^{\frac{3}{2}}}.$$

To find the matter density ρ in the radiation era, we use (7.1.5) and (8.1.4) to get

$$\rho(t) \overset{(t<t_{\text{E}})}{=} \rho(t_{\text{E}}) \frac{R^3(t_{\text{E}})}{R^3(t)} = \frac{\rho(t_{\text{E}})\,T_{\text{r}}^3(t)}{T_{\text{r}}^3(t_{\text{E}})} = \frac{\rho_{\text{r}}(t_{\text{E}})\,T_{\text{r}}^3(t)}{T_{\text{r}}^3(t_{\text{E}})} = \frac{aT_{\text{r}}(t_{\text{E}})\,T_{\text{r}}^3(t)}{c^2},$$

so that M_{J} becomes

▶ $$M_{\text{J}} = \frac{c^4 T_{\text{r}}(t_{\text{E}})}{G^{\frac{3}{2}}a^{\frac{1}{2}}T_{\text{r}}^3}. \qquad (8.2.3)$$

As the universe expands and the radiation cools, M_J increases (as T_r^{-3}) and its maximum value, at the close of the radiation era, is

$$\blacktriangleright \quad M_J^{\binom{\text{radiation}}{\text{dominated}}(t=t_E)} = \frac{c^4}{G^{\frac{3}{2}}a^{\frac{1}{2}}T_r^2(t_E)} \approx 6 \times 10^{48} \text{ kg} \approx 3 \times 10^{18} \, M_\odot, \quad (8.2.4)$$

where we have taken $T_r(t_E)$ as 2000 K. This value for M_J greatly exceeds known galactic masses. Thus an inhomogeneity of galactic size – a 'proto-galaxy' – would, after an initial unstable contracting phase, spend most of the radiation era oscillating acoustically.

In the *matter-dominated era* ($t > t_E$), the radiation density is negligible, and $\rho_{\text{tot}} \approx \rho$. We can treat the matter as an ideal gas of monatomic hydrogen, with specific heat ratio $\gamma = \frac{5}{3}$ and an equation of state

$$p = \frac{\rho k T_m}{m_H}.$$

where m_H is the mass of a hydrogen atom and T_m is the matter temperature, which is not the same as T_r. Thus the speed of sound is

$$v = \sqrt{\frac{dp}{d\rho}} = \sqrt{\frac{kT_m}{m_H}}.$$

(assuming isothermal oscillations for simplicity) and M_J is, from (8.2.1),

$$M_J = \frac{4\pi}{3} \left(\frac{kT_m}{Gm_H}\right)^{\frac{3}{2}} \frac{1}{\rho^{\frac{1}{2}}}.$$

It is convenient to express T_m and ρ in terms of the radiation temperature T_r: first we realise that the expansion of the matter is adiabatic, so

$$T_m \times \text{volume}^{\gamma-1} = \text{constant},$$

i.e. $\quad \dfrac{T_m}{\rho^{\gamma-1}} = \dfrac{T_m}{\rho^{\frac{2}{3}}} = \text{constant} = \dfrac{T_r(t_E)}{\rho(t_E)^{\frac{2}{3}}}$

(since $T_r = T_m$ by definition at $t = t_E$), and

$$T_m = T_r(t_E) \left(\frac{\rho}{\rho(t_E)}\right)^{\frac{2}{3}}.$$

Now we recall that $T_r \propto R^{-1}$, while $\rho \propto R^{-3}$, so that

$$\rho(t) = \rho(t_E) \left[\frac{T_r(t)}{T_r(t_E)}\right]^3.$$

Thus, finally, M_J becomes

$$\blacktriangleright \quad M_J = \frac{4\pi}{3} \left(\frac{kT_r}{Gm_H}\right)^{\frac{3}{2}} \frac{1}{\rho(t_E)^{\frac{1}{2}}}. \quad (8.2.5)$$

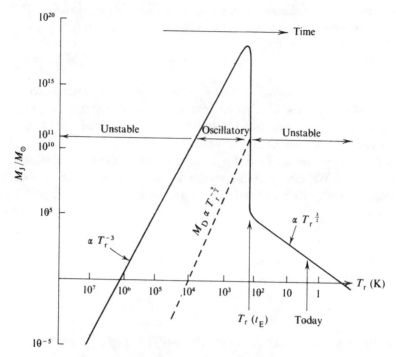

Figure 62. History of Jeans mass M_J, and the minimum mass M_D undamped by photon viscosity, as the cosmic fluid evolves.

As the matter-dominated universe expands and cools, M_J decreases (as $T_r^{\frac{3}{2}}$) and its maximum value, at the beginning of the matter era, is

▶
$$M_J^{\binom{\text{matter}}{\text{dominated}}}{}^{(t=t_E)} = \frac{4\pi}{3}\left(\frac{kT_r(t_E)}{Gm_H}\right)^{\frac{3}{2}}\frac{1}{\rho^{\frac{1}{2}}(t_E)} \approx 6\times 10^{35}\text{ kg} \approx 3\times 10^5\,M_\odot,$$

(8.2.6)

where we have evaluated $\rho(t_E)$ using the conservation law (7.1.5), the contraction factor $R(t_E)/R(t_0) = 10^{-3}$ and the present mass density $\rho_0 = 3\times 10^{-28}$ kg m^{-3}. This value for M_J is much less than a typical galactic mass, so that any protogalaxies surviving until t_E would contract freely thereafter. The behaviour of the Jeans mass as a function of T_r (which in the early universe is proportional to $t^{-\frac{1}{2}}$, from equation (8.1.12)) is shown in figure 62.

We have one more effect to consider, damping from *photon viscosity* (figure 61(c)) in the radiation era. This arises as follows: the close coupling between matter and radiation before t_E is maintained by collisions between photons and electrons (protons are much heavier so that collisions with them involve much smaller energy transfers). In the presence of inhomo-geneities, however, the equilibrium established by these 'Compton scatter-ings' is not perfect, and the radiation will iron out a matter inhomogeneity of (linear) dimension X in roughly the time it takes a photon to diffuse several times X (say $5X$). Thus when the universe is of age $t(< t_E)$ all inhomogeneities smaller than $X(t)$, where $X(t)$ is one-fifth of the dis-tance diffused by a photon in the time t, will have been damped out. Because diffusion is a 'random walk' process, this distance is $L\sqrt{N}$, where L is the photon mean free path between collisions and N the total number of collisions. Photons travel with speed c, so that

$$X(t) \approx \tfrac{1}{5}L\sqrt{N} = \tfrac{1}{5}L\sqrt{(ct/L)} = \tfrac{1}{5}\sqrt{(Lct)}.$$

From kinetic theory,

$$L = 1/n_e(t)\,\sigma = m_H/\rho(t)\,\sigma,$$

where n_e is the number density of scatterers (i.e. electrons) at t, and σ the scattering cross-section, which for photon-electron collisions has the 'Thomson' value

$$\sigma = e^4/6\pi\epsilon_0^2 m_e^2 c^4 = 6.65 \times 10^{-29}\ \mathrm{m}^2.$$

Thus the mass $M_D(t)$ of the smallest inhomogeneity that has survived undamped until the time t is

▶ $$M_D(t) = \frac{4\pi\rho(t)}{3}\left(\frac{X(t)}{2}\right)^3 = \frac{4\pi}{3\times 10^3\rho(t)^{\frac{1}{2}}}\left(\frac{m_H\,ct}{\sigma}\right)^{\frac{3}{2}}. \tag{8.2.7}$$

As t increases, larger inhomogeneity masses $M(t)$ get damped out, until at the close of the radiation era the minimum undamped mass is

▶ $$M_D(t_E) = \frac{4\pi}{3\times 10^3\rho(t_E)^{\frac{1}{2}}}\left(\frac{m_H\,ct_E}{\sigma}\right)^{\frac{3}{2}} \approx 2\times 10^{42}\ \mathrm{kg} \approx 10^{12}M_\odot. \tag{8.2.8}$$

This is encouragingly close to the mass of typical large galaxies; however the discovery of any 'missing mass' would greatly reduce the value of $M_D(t_E)$ (if ρ_0 is increased by a factor α, $M_D(t_E)$ is reduced by a factor α^7). As a function of temperature, $M_D(t)$ varies as $T_r^{-\frac{9}{2}}$, and this is shown on figure 62. After t_E photon viscous damping ceases, because recombination has occurred and the matter is transparent to radiation.

We can sum up the results of the theory as follows: all primaeval in-homogeneities with masses less than that of a typical galaxy will be damped out during the radiation era. Larger masses will survive into the matter era and then contract freely under gravity. These conclusions are confirmed by more thorough calculations which take account of the possibility that recombination may not occur at the time t_E when ρ and ρ_r

are equal. The theory only gives us a *minimum* undamped mass; why do we not observe galaxies with masses much larger than $10^{12}\,M_\odot$? Probably because for any reasonable distribution of primaeval inhomogeneities the density irregularities would be weaker the larger the scale, so that galaxies with masses greatly in excess of $10^{12}M_\odot$ would be not only very rare but also so slow to contract that they might not yet have condensed.

Now we turn to the theory of the *forms* of galaxies (see the frontispiece). It is quite easy to explain why they are often pancake-shaped rather than spherical. We need to assume only that in these cases the original irregularity possessed *angular momentum* about some axis, z, say. As the galaxy contracts under its self-gravitation, this angular momentum is *conserved*, since there are no external torques on the galaxy. In order for such a spinning object to contract towards the z axis, work must be done on it. To see this, consider a star of mass m near the edge of the galaxy; assume that at time t it moves in a circular orbit of radius r with speed v. Thus its angular momentum L and kinetic energy K are

$$L = mrv, \quad K = mv^2/2 = L^2/2mr^2.$$

As the contraction proceeds, r decreases while L remains constant; therefore K increases – angular momentum acts like a repulsive force resisting contraction. This increase of K requires work, which is provided by the gravitational potential

$$V_{\text{grav}} = -GmM_g/r$$

where M_g is the mass of the galaxy.

Now, the required energy increases as r^{-2}, but the available gravitational energy increases only as r^{-1}, so that there must come a radius r_g at which gravity can no longer fuel the contraction. The value of r is obtained from the condition for $K+V_{\text{grav}}$ to be an extremum, i.e. for there to be no radial force. Thus

$$\frac{d}{dr}\left(\frac{L^2}{2mr^2} - \frac{GM_g m}{r}\right) = 0 \quad \text{if} \quad r = r_g,$$

or

$$-\frac{L^2}{mr_g^3} + \frac{GM_g m}{r_g^2} = 0,$$

i.e.

▶ $$r_g = L^2/GM_g m^2. \tag{8.2.9}$$

But $L = mv_g r_g$, where v_g is the speed of a star on the edge of a contracted galaxy. If we insert this formula for L into (8.2.9), we simply obtain again the formula (2.2.9) that was used to obtain M_g from measurements of the rotation rate. Now we take $L = \mathscr{L}/N$, $m = M_\odot$ and $M_g = NM_\odot$, where $\mathscr{L}$ is the angular momentum of the whole galaxy and N is the number of stars in the galaxy; (8.2.9) becomes

▶ $$r_g = \mathscr{L}^2/GM^3_\odot N^3. \tag{8.2.10}$$

This tells us the final radius of a galaxy that condensed from a fluctuation of mass $NM_\odot$ and *total* angular momentum $\mathscr{L}$.

There is, however, no such 'centrifugal barrier' to limit contraction *along* the z axis, so that this part of the collapse is almost complete. Therefore galaxies resulting from fluctuations possessing angular momentum should be disc-shaped, and this is often observed.

What about the spiral arms of galaxies? These are not well understood. Looking at spirals shown in the frontispiece, it seems likely that the arms are associated with some kind of swirling or vortex motion. Now vortices are the elementary constituents of *turbulence* and there have been many cosmologies (starting with those of Kepler and Descartes) in which the universe has been regarded as analogous to a turbulent fluid. However, immense mathematical difficulties attend the development of such theories, for the following reasons: at the transition time t_E the speed of sound drops abruptly from $c/\sqrt{3}$ to the much smaller value $\sqrt{(kT_E/m_H)} \sim 4 \times 10^3$ m s^{-1}. Thus turbulent eddies which were subsonic in the radiation era can be supersonic in the matter era. Unfortunately the theory of supersonic turbulence is in its infancy, and its rigorous application to the development of galactic structure lies in the future. However, one promising area of subsonic turbulence theory is that concerned with the development of irregularities: smaller eddies tend to capture energy from, and grow at the expense of, larger eddies and this might be another reason why there are no galaxies much larger than $10^{12}M_\odot$.

8.3 Ylem

Before the time $t_E \sim 10^6$ years, the universe was radiation dominated and the history of the matter was governed largely by the fireball of photons. If the reasonable assumption is made that only elementary particles were present in the original ylem, it follows that the nuclei, atoms, etc. that now exist must have evolved later. We have already seen how the fireball temperature T_r dropped below 4000 K at about t_E, thus permitting electrons and protons to combine into atomic hydrogen. It is currently believed that the formation of *nuclei* began at a much earlier date. Starting with hydrogen, the first step is to understand the synthesis of *helium*.

This can occur as a result of a variety of nuclear reactions, such as

$$p + n \rightarrow D \; (= \text{deuterium} = {}^2H),$$

and $D + n \rightarrow T \; (= \text{tritium} = {}^3H), \; T + p$

or $D + p \rightarrow {}^3He, \qquad\qquad {}^3He + n \longrightarrow {}^4He + \text{energy}$

or $D + D \rightarrow n + {}^3He \rightarrow T + p$

Deuterium is an essential ingredient in the production of significant amounts of helium. However, the binding energy of deuterium is 2.2 MeV, and if the thermal energy kT_r of the fireball photons is greater than this, any deuterons will be rapidly photodissociated. Thus helium cannot be produced until the temperature falls below

$$T_r = 2.2 \text{ MeV}/k \sim 10^{10} \text{ K.}$$

This occurs at a time t given by the dynamical equation (8.1.12), which we can rewrite as

$$\blacktriangleright \qquad t = \left(\frac{3}{32\pi Ga}\right)^{\frac{1}{2}} \frac{c}{T_r^2} = 2.3 \left(\frac{10^{10}}{T_r}\right)^2 \text{s,} \qquad (8.3.1)$$

where T_r is in kelvins. Thus it was possible for nucleosynthesis to begin when the universe was two seconds old.

Before this time, the photon energy kT_r exceeded about 1 MeV. Now, this is just the energy required for electron–positron pair production (the rest mass energy of the electron is 0.511 MeV). Therefore until $t \sim 2$ s the fireball was flooded with electrons and positrons. In the presence of these particles, neutrons and protons can transform into one another, and in the quasi-equilibrium then existing the ratio r of neutrons to protons is governed by their difference in rest mass energy $\Delta mc^2 = 1.294$ MeV, being given by the Boltzmann factor

$$\blacktriangleright \qquad r = \exp\left(-\Delta mc^2/kT_r\right) = 0.22 \quad \text{if} \quad T_r = 10^{10} \text{ K.} \qquad (8.3.2)$$

Two neutrons and two protons are required to produce one ${}^4He$ nucleus. If the nucleosynthesis is complete, all the neutrons are used up, so that the number of ${}^4He$ nuclei is proportional to $r/2$. The number of protons (${}^1H$) left over is proportional to $1 - r$. The mass of ${}^4He$ is four times greater than that of ${}^1H$, so the final *mass abundance of helium* is predicted by this simple argument to be

$$\blacktriangleright \qquad \frac{{}^4He}{{}^4He + {}^1H} = \frac{4 \times r/2}{(4 \times r/2) + [1 \times (1-r)]} = \frac{2r}{1+r} = 36 \text{ per cent.} \qquad (8.3.3)$$

More precise calculations have been carried out, taking into account all the different fusion reactions leading to ^{4}He, and also the changing rates of deuteron photodisintegration and electron–positron pair production as the fireball expands and cools; it turns out that most of the synthesis occurs at the slightly lower temperature of 10^9 K, when the universe is about 200 s old. The result is a predicted helium mass abundance of 27 per cent. A variety of observational evidence indicates that the actual cosmic abundance is about 25 per cent, but the interpretation of the experiments is complicated and this figure is by no means accepted by all cosmologists. If the 'helium problem' is solved, and the value of 25 per cent confirmed, this will represent a triumph for the 'hot big bang' theory.

What about heavier elements? It is difficult to see how fusion in the fireball could have produced significant numbers of nuclei heavier than ^{4}He, because there are no stable 'intermediate' nuclei with atomic weights 5 and 8. Therefore it is generally believed that the small present-day cosmic abundance of heavier elements originated much later than helium, as a result of hydrogen burning in *stars*. Of course this also produces more helium, but the amount is estimated to be small compared with that produced in the fireball.

Now we examine the ylem in the first moments of the big bang. We have seen that before $t \sim 2$ s electrons and positrons existed in great numbers along with the photons. At even earlier times the temperature was high enough to produce heavier particle–antiparticle pairs. For example, muons have rest mass energy of about 100 MeV, and so $\mu^+ - \mu^-$ pairs existed before

$$T_r = \text{energy}/k \sim 10^{12} \ K, t \sim 2 \times 10^{-4} \text{ s}.$$

Protons and antiprotons, whose rest energies are about 10^3 MeV, existed in abundance before

$$T_r \sim 10^{13} \ K, t \sim 2 \times 10^{-6} \text{ s}.$$

In those first moments, the number of elementary particles was vastly greater than it is now, and the numbers of particles and antiparticles were almost equal. Almost but not quite. If the numbers had been *precisely* equal, the subsequent annihilation would surely have been almost complete, and there would be virtually no matter today. It has been calculated on this 'symmetrical hot big bang model' that the present number density of nucleons would be only about 10^{-18} that of photons, whereas observation indicates a value in excess of 10^{-10}, and this requires an initial fractional excess of 10^{-8} in the number of nucleons over antinucleons. However, this poses a problem: why was a small asymmetry between matter and antimatter built into the original ylem? We do not know the answer. It has been suggested that the amounts of matter and antimatter were in fact equal, and that what remains today are pockets of matter and antimatter that somehow became separated and escaped the various early stages of annihilation. Various segregation mechanisms have been proposed and it is conceivable that the matter and antimatter became separated into regions that later condensed into galaxies and antigalaxies. Observationally, no antinucleons have been detected in primary cosmic radiation, and there is only a very small flux of γ rays of the type expected to be produced by annihilation in regions where matter and antimatter are mixed.

Finally we consider the very beginning, $t = 0$. The relativistic equations (8.1.13) and (8.1.14) tell us that the universe began with a singularity of infinite density ρ_r and zero cosmic scale factor R. The nature of this singularity is different for closed universes ($k = +1$) than for flat ($k = 0$) or open ($k = -1$) universes. If $k = +1$, there is a finite amount of matter and radiation packed into zero initial proper volume; this 'point', however, includes the whole of space – there is nothing 'outside'. If k is 0 or -1, the total amount of matter and radiation is infinite, as is the proper volume, and the initial singularity is therefore 'everywhere'. Now, the initial singularity is a consequence of the equations of general relativity, and this theory might break down under conditions of very high density and pressure. The most likely possibility is that *quantum effects* might occur, and prevent localisation within infinitely small volumes. (Remember that the first triumph of quantum mechanics was to solve the problem of the electromagnetic collapse of the atom that would be caused by classically radiating electrons spiralling into the nucleus.) Let us estimate the time t_Q at which these quantum effects would occur. The expansion of the universe is described by Hubble's constant, which for the radiation-dominated era is given by (8.1.14) as

$$H(t) = \frac{\dot{R}(t)}{R(t)} = \frac{1}{2t}.$$

This rate cannot exceed the rate of oscillation ω of the wave functions of elementary particles at time t; ω is given by

$$\hbar\omega = \text{energy} = kT,$$

i.e. $$\omega = \frac{kT_r}{\hbar} = \frac{k}{\hbar}\left(\frac{3c^2}{32\pi Ga}\right)^{\frac{1}{4}}\frac{1}{\sqrt{t}},$$

where we have used (8.1.12). General relativity is inapplicable when

Table 4

Time t(s)	What happens	
0		
$10^{-43} = t_Q$	Quantum chaos	
2×10^{-6}	Nucleons annihilate	
2×10^{-4}	Muons annihilate	Radiation-dominated era
2	Electron–positron pairs annihilate	
200	Helium nucleosynthesis	
$10^{14} \sim t_E$	Hydrogen atoms form	Matter-dominated era
$10^{18} \sim t_o$	Today	

$\omega < H(t)$, that is, when $t < t_Q$ and $T_r > T_{r, Q}$, where

$$\blacktriangleright \qquad t_Q = \frac{\hbar^2}{4k^2} \left(\frac{32\pi Ga}{3c^2} \right)^{\frac{1}{2}} = \pi \left(\frac{hG}{45c^5} \right)^{\frac{1}{2}} \sim 10^{-43} \text{ s,}$$

$$T_{r, Q} \sim 5 \times 10^{31} \text{ K.} \qquad\qquad (8.3.4)$$

If this argument is correct, the universe has expanded from a *finite R*, given by

$$\blacktriangleright \qquad R(t_Q) = \frac{R(t_0) \, T_r(t_0)}{T_{r, Q}} \sim 10^{-21} \text{ light-years} \sim 10^{-5} \text{ m.} \qquad (8.3.5)$$

The history of the matter in the universe is summarised in table 4.

In this section we have made the ultimate extrapolations of our physical laws. Is the resulting world-picture so much 'science fiction', as Brillouin believed, or should we echo the words of W. S. Gilbert in 'The Mikado' (1885):

> 'I am, in point of fact, a particularly haughty and exclusive person of pre-Adamite ancestral descent... I can trace my ancestry back to a protoplasmal primordial atomic globule.'

APPENDIX A

Labelling astronomical objects

The direction of objects in the sky is determined by two angles, declination and right ascension. If we imagine all the objects projected onto the surface of a sphere ('the celestial sphere') then these two angles are analogous to latitude and longitude, which determine position on the Earth's surface. The celestial sphere is divided into constellations ('con-stellations'), which are analogous to countries on the Earth. Constellations contain prominent groupings of stars such as the Great Bear (Ursa Major) or Orion; these stars may have no real dynamical connection with one another, in that their distances from us may differ enormously. Within constellations, objects are labelled with a Greek letter α, β...ω roughly according to their brightness, and, for fainter objects, by a number 1, 2.... Thus '61 Cygni' means 'Object number 61 in the constellation Cygnus (the swan)'. Galaxies, and other objects of non-stellar appearance such as nebulae (gas clouds) within our own galaxy, have a separate nomenclature: thus the closest galaxy to ours, in the constellation of Andromeda is 'M 31' – the 31st object in the old catalogue of Messier – or, alternatively, 'NGC 224' – the 224th object in the *New General Catalogue*. In the catalogues all these objects are of course precisely located by their 'map reference', that is, by their right ascension and declination.

APPENDIX B

Theorema egregium

This means 'the very beautiful theorem', and is the title given by Gauss to the derivation of his curvature formula (4.3.5). This formula lies right at the heart of the differential geometry of surfaces, and is also the basis of our simplified treatment of general relativity. A full and rigorous derivation would be very lengthy, so we shall present the outline of the argument. For general orthogonal coordinates x^1 and x^2 on the surface, the metric distance formula (4.1.7) becomes

$$\Delta s^2 = g_{11}(x^1, x^2)\,(\Delta x)^2 + g_{22}(x^1, x^2)\,(\Delta x^2)^2. \tag{B.1}$$

What we have to do is calculate the curvature K at any point P on the surface in terms of derivatives of g_{11} and g_{22}.

K is defined by equation (4.3.3) in terms of the circumference C of a small geodesic circle centred on P. We begin by introducing *geodesic polar coordinates* r, ϕ defined as follows: draw the geodesics emanating from P, and label them by the angle ϕ which they make with the positive x^1 axis at P; thus $0 \leqslant \phi < 2\pi$. Distances from P along the geodesics constitute the 'radial' coordinate r. The locus of points with the same value of r is a geodesic circle, and it can be shown that all geodesic circles are orthogonal to the geodesics they cut. Thus the metric in these coordinates is

$$\Delta s^2 = \Delta r^2 + r^2 f(r, \phi)\,\Delta\phi^2, \tag{B.2}$$

where $f(r, \phi)$ is a function whose form depends on the surface, and whose value at P must be unity; so that r and ϕ reduce to ordinary polar coordinates in the infinitesimal flat neighbourhood of P. Around a geodesic circle of radius a, ϕ varies from 0 to 2π while r maintains the constant value a. Therefore the circumference is

$$C = \int \Delta s = a \int_0^{2\pi} (\sqrt{f(a, \phi)})\,d\phi. \tag{B.3}$$

We shall see that $f(a, \phi)$ can be expanded about P as

$$f(a, \phi) = 1 + a^2 q + \dots, \tag{B.4}$$

[160]

where q is constant. Therefore

$$C = a \int_0^{2\pi} (1 + a^2 q/2 + \dots) \, d\phi = 2\pi a + \pi a^3 q + \dots, \tag{B.5}$$

so that from (4.3.3) the curvature is

$$K = -3q. \tag{B.6}$$

It is now necessary to express q in terms of g_{11} and g_{22}, by transforming the expression (B.1) for Δs^2 in terms of x^1 and x^2 into the expression (B.2) involving r and ϕ. As a first step, we define *quasicartesian* coordinates ξ and η near P, by the relations

$$\xi \equiv (\sqrt{g_{11}(x_P^1, x_P^2)}) (x^1 - x_P^1), \quad \eta \equiv (\sqrt{g_{22}(x_P^1, x_P^2)}) (x^2 - x_P^2), \tag{B.7}$$

where x_P^1, x_P^2 are the coordinates of P. Then

$$\Delta s^2 = \gamma_1(\xi, \eta) \, \Delta \xi^2 + \gamma_2(\xi, \eta) \, \Delta \eta^2, \tag{B.8}$$

where

$$\left. \begin{aligned} \gamma_1(\xi, \eta) &= g_{11}(x^1, x^2)/g_{11}(x_P^1, x_P^2); \\ \gamma_2(\xi, \eta) &= g_{22}(x^1, x^2)/g_{22}(x_P^1, x_P^2); \\ \gamma_1(0, 0) &= \gamma_2(0, 0) = 1. \end{aligned} \right\} \tag{B.9}$$

ξ and η are unknown functions of r and ϕ; to find them we write (B.8) as

$$\Delta s^2 = \gamma_1(\xi, \eta) \left[\frac{\partial \xi}{\partial r} \Delta r + \frac{\partial \xi}{\partial \phi} \Delta \phi \right]^2 + \gamma_2(\xi, \eta) \left[\frac{\partial \eta}{\partial r} \Delta r + \frac{\partial \eta}{\partial \phi} \Delta \phi \right]^2 \tag{B.10}$$

and equate the coefficients of Δr^2, $\Delta \phi \, \Delta r$ and $\Delta \phi^2$ with those in (B.2). This gives

$$\left. \begin{aligned} 1 &= \gamma_1(\xi, \eta) \left(\frac{\partial \xi}{\partial r} \right)^2 + \gamma_2(\xi, \eta) \left(\frac{\partial \eta}{\partial r} \right)^2, \quad &a \\ 0 &= \gamma_1(\xi, \eta) \frac{\partial \xi}{\partial r} \frac{\partial \xi}{\partial \phi} + \gamma_2(\xi, \eta) \frac{\partial \eta}{\partial r} \frac{\partial \eta}{\partial \phi}, \quad &b \\ r^2 f(r, \phi) &= \gamma_1(\xi, \eta) \left(\frac{\partial \xi}{\partial \phi} \right) + \gamma_2(\xi, \eta) \left(\frac{\partial \eta}{\partial \phi} \right)^2. \quad &c \end{aligned} \right\} \tag{B.11}$$

The first two equations determine $\xi(r, \phi)$ and $\eta(r, \phi)$ and the third gives $f(r, \phi)$ and hence K, according to (B.4) and (B.6). We only require ξ and η near P, so we write

$$\left. \begin{aligned} \xi &\equiv r \cos \phi + \tfrac{1}{2} r^2 \psi_1(\phi) + \tfrac{1}{3} r^3 \chi_1(\phi) + \dots \\ \eta &\equiv r \sin \phi + \tfrac{1}{2} r^2 \psi_2(\phi) + \tfrac{1}{3} r^3 \chi_2(\phi) + \dots. \end{aligned} \right\} \tag{B.12}$$

We also expand the metric functions $\gamma_i(\xi, \eta)$ (where $i = 1, 2$), in the form

$$\gamma_i(\xi, \eta) = 1 + r A_i(\phi) + r^2 B_i(\phi) + \dots, \tag{B.13}$$

where

$$A_i(\phi) = \cos\phi\gamma_{i\xi} + \sin\phi\gamma_{i\eta},$$

$$\left.\begin{array}{l} B_i(\phi) = \tfrac{1}{2}(\psi_1(\phi)\,\gamma_{i\xi} + \psi_2(\phi)\,\gamma_{i\eta} + \cos^2\phi\gamma_{i\xi\xi} + \sin^2\phi\gamma_{i\eta\eta} \\ \qquad\qquad +2\sin\phi\cos\phi\gamma_{i\xi\eta}), \end{array}\right\} \quad \text{(B.14)}$$

and where we have introduced the shorthand notation

$$\gamma_{i\xi} \equiv \frac{\partial\gamma_i(0,0)}{\partial\xi}\ \text{etc.} \tag{B.15}$$

If we insert (B.12) and (B.13) into (B.11a) and (B.11b), we find that the terms not involving r cancel. Equating to zero the terms in r and r^2, we get from (B.11a),

$$\left.\begin{array}{l} 0 = A_1\cos^2\phi + 2\psi_1\cos\phi + A_2\sin^2\phi + 2\psi_2\sin\phi, \qquad a \\[4pt] 0 = B_1\cos^2\phi + 2\psi_1 A_1\cos\phi + \psi_1^2 + 2\chi_1\cos\phi + B_2\sin^2\phi \\ \qquad\quad +2\psi_2 A_2\sin\phi + \psi_2^2 + 2\chi_2\sin\phi. \quad b \end{array}\right\} \quad \text{(B.16)}$$

From (B.11b) we get

$$\left.\begin{array}{l} 0 = \tfrac{1}{2}\cos\phi\,\psi_1' - \sin\phi\,\psi_1 - A_1\sin\phi\cos\phi + \tfrac{1}{2}\sin\phi\psi_2' \\ \qquad\qquad +\cos\phi\psi_2 + A_2\sin\phi\cos\phi, \quad a \\[6pt] 0 = \tfrac{1}{3}\cos\phi\chi_1' - \sin\phi\chi_1 - B_1\sin\phi\cos\phi - \sin\phi A_1\psi_1 \\ \qquad +\tfrac{1}{2}\cos\phi\psi_1' A_1 + \tfrac{1}{2}\psi_1'\psi_1 + \tfrac{1}{3}\sin\phi\chi_2' + \cos\phi\chi_2 \\ \qquad + B_2\sin\phi\cos\phi + \cos\phi A_2\psi_2 + \tfrac{1}{2}\sin\phi\psi_2' A_2 + \tfrac{1}{2}\psi_2'\psi_2. \quad b \end{array}\right\} \quad \text{(B.17)}$$

(Primes denote differentiation with respect to ϕ.) By judiciously differentiating and substituting among these equations, the functions $\psi_1(\phi)$, $\psi_2(\phi)$, $\chi_1(\phi)$ and $\chi_2(\phi)$ involved in the expansion (B.12) of ξ and η can be obtained.

Knowing ξ and η, the next step is to employ (B.11c) to find $f(r,\phi)$; we use the expansion

$$f(r,\phi) \equiv 1 + rp(\phi) + r^2q(\phi) + \ldots, \tag{B.18}$$

and equate powers of r in (B.11c). The terms in r^2 cancel identically, while those in r^3 and r^4 give

$$p(\phi) = -\sin\phi\psi_1' + \sin^2\phi A_1 + \cos\phi\psi_2' + \cos^2\phi A_2, \tag{B.19}$$

$$\begin{array}{l} q(\phi) = \tfrac{1}{4}\psi_1'^2 - \tfrac{2}{3}\chi_1'\sin\phi - \tfrac{1}{3}A_1\psi_1'\sin\phi + B_1\sin^2\phi \\ \qquad\quad +\tfrac{1}{4}\psi_2'^2 + \tfrac{2}{3}\chi_2'\cos\phi + A_2\psi_2'\cos\phi + B_2\cos^2\phi. \end{array} \tag{B.20}$$

The solution of equations (B.17) and substitution into (B.19) and (B.20) involves extremely heavy algebra. Considerable simplification results from choosing a particular angle, say $\phi = 0$, at which to calculate ψ_1, ψ_2, χ_1, χ_2 and hence p and q. Even so, the working covers several sheets of paper. For

p, we obtain the value zero. This result contains no reference to the special property of $\phi = 0$, namely that this direction corresponds to the positive ξ axis; therefore we must have, for all angles,

$$p(\phi) = 0. \tag{B.21}$$

For q, we obtain the value

$$q = \tfrac{1}{6}\gamma_{2\xi\xi} + \tfrac{1}{6}\gamma_{1\eta\eta} - \tfrac{1}{12}\gamma_{1\eta}^2 - \tfrac{1}{12}\gamma_{2\xi}^2 - \tfrac{1}{12}\gamma_{1\xi}\gamma_{2\xi} - \tfrac{1}{12}\gamma_{1\eta}\gamma_{2\eta}. \tag{B.22}$$

This contains ξ and η symmetrically, and so must also hold for all ϕ. These results justify the expansion (B.4).

To find the curvature, we use (B.6), and obtain

$$K = -\tfrac{1}{2}(\gamma_{2\xi\xi} + \gamma_{1\eta\eta}) + \tfrac{1}{4}(\gamma_{1\eta}^2 + \gamma_{2\xi}^2 + \gamma_{1\xi}\gamma_{2\xi} + \gamma_{1\eta}\gamma_{2\eta}). \tag{B.23}$$

The final step is to transform from γ_1, γ_2, ξ, η back to g_{11}, g_{22}, x^1 and x^2 using (B.7); a typical term is

$$\gamma_{2\xi\xi} \equiv \frac{\partial^2 \gamma_2}{\partial \xi^2} = \frac{\partial^2 (g_{22}(x^1, x^2)/g^{22}(x_P^1, x_P^2))}{\partial (x^1)^2 \times g_{11}(x_P^1, x_P^2)} = \left[\frac{1}{g_{11}g_{22}} \frac{\partial^2 g_{22}}{\partial (x^1)^2} \right]_{\substack{x^1 = x_P^1 \\ x^2 = x_P^2}}$$

$$\tag{B.24}$$

The other terms in (B.23) transform similarly, and the final result is precisely the Gauss curvature formula (4.3.5), namely

$$K = \frac{1}{2g_{11}g_{22}} \left\{ -\frac{\partial^2 g_{11}}{\partial (x^2)^2} - \frac{\partial^2 g_{22}}{\partial (x^1)^2} + \frac{1}{2g_{11}} \left[\frac{\partial g_{11}}{\partial x^1} \frac{\partial g_{22}}{\partial x^1} + \left(\frac{\partial g_{11}}{\partial x^2} \right)^2 \right] \right.$$

$$\left. + \frac{1}{2g_{22}} \left[\frac{\partial g_{11}}{\partial x^2} \frac{\partial g_{22}}{\partial x^2} + \left(\frac{\partial g_{22}}{\partial x^1} \right)^2 \right] \right\}.$$

$$\tag{B.25}$$

Problems

1. Show how the mean distance from the Earth to the Sun, $r_\oplus$, can be calculated from a knowledge of the orbital periods $T_\oplus$ and $T_{\vec\delta}$ of the Earth and Mars, and the distance R between the Earth and Mars at their closest approach. (Assume circular orbits.)

2. What is the greatest distance that could be determined using the parallax method with the world's largest telescope (mirror diameter 200 in)?

3. A star has absolute magnitude M. What is its absolute luminosity L?

4. The world's largest telescope can just detect an object whose apparent magnitude m is 22.7. If the Sun were on the far edge of the galaxy, could we detect it? (Neglect absorption.)

5. Consider an idealised galaxy where most of the mass is concentrated within a spherical core of uniform density. Show that the orbital velocity $v(r)$ of a star at a distance r from the centre is (a) $v(r)$ proportional to r if r is inside the core, (b) $v(r)$ proportional to $r^{-\frac{1}{2}}$ if r is outside the core.

6. Estimate the gravitational red shift z of light escaping from a galaxy after being emitted from a star at the 'surface' of the core (~ 1000 pc from the centre). Assume that this core contains 10^9 'Suns'.

7. At what rate should hydrogen atoms be created per cubic metre in order that the density of the expanding universe could remain constant at its present value $\rho \sim 10^{-28}$ kg m^{-3}?

8. What fractional charge difference between the electron and the proton would make the electrostatic repulsion between the Earth and the Moon cancel the gravitational attraction between these bodies? Obtain a rough estimate only by assuming that the Earth and Moon are made of hydrogen.

9. If the inertial mass m_i were not the same as the gravitational mass m_g, what would be the period T of a simple pendulum of length l at a distance r from the centre of a spherical mass M_g?

10. Imagine that mass could be negative as well as positive. Discuss the motion of two objects, initially at rest and released to respond to their mutual gravitation, if (a) the masses are equal and positive, (b) the masses are equal and negative, (c) the masses are equal and opposite. Does case (c) violate the law of conservation of momentum?

11. A body of mass m is acted on by a body of mass M at a distance r. M has an acceleration a relative to m. In order to satisfy Mach's principle, the force exerted by M on m must contain a part F proportional to ma. Using dimensional analysis, derive the simplest possible formula for the magnitude F of F.

12. A satellite moves in a circular orbit around the Earth. Inside the satellite, two objects are simultaneously released from rest relative to the satellite, one object being Δr above the other. After one revolution, the objects will no longer lie on the same vertical. What is their relative horizontal displacement? (This indicates the departure of a 'laboratory' of finite size from an ideal local inertial frame.)

13. Light is emitted horizontally *in vacuo* near the Earth's surface, and falls freely under the action of gravity. Through what vertical distance has it fallen after travelling 1 km?

14. By direct transformation from Cartesian coordinates, calculate the metric tensor $g_{\mu\nu}$ in a flat three-dimensional space in terms of (a) spherical polar coordinates, (b) cylindrical coordinates.

15. Show that if two neighbouring events have a space-like separation, the straight world line joining them corresponds to the longest proper distance between the events.

16. How can a body pursue a straight path in space and yet have a curved world line?

17. Show that the angle sum of a spherical triangle lies between 180° and 900°.

18. Show that the curvature K of a sphere is given in terms of the area A of circles of radius a by

$$K = (12/\pi) \lim_{a \to 0} [(\pi a^2 - A)/a^4].$$

19. Using Gauss's curvature formula (4.3.5), show that $K = 0$ for a plane when plane polar coordinates are used, and that $K = 1/R^2$ for a sphere of radius a when spherical polar coordinates are used.

20. Calculate the curvature $K(r)$ of a surface whose metric distance formula is
$\Delta s^2 = f(r)\,\Delta r^2 + r^2 \Delta\phi^2$.

21. Calculate the circumference c and volume V of a hypersphere of proper radius a in a three-dimensional space of constant curvature K.

22. An isolated mass M lies at the spatial origin of a frame of reference. The curvature K of a geodesic spacelike surface passing through the origin, is, at a point whose radial coordinate is r,

    $K = qMG^i c^j r^k$,

    where q is dimensionless. Find the indices i, j and k.

23. What is the Schwarzschild radius of (a) a galaxy (in parsecs), (b) a proton (in metres)?

24. Calculate according to Newtonian mechanics the escape velocity from the surface of the Schwarzschild sphere surrounding a massive body.

25. Calculate the radial coordinate at which light travels in a circular path round a body of mass M, using (a) Newtonian mechanics, (b) general relativity. Express your answers as a multiple of the Schwarzschild radius.

26. Prove the result (5.5.6).

27. Show that the present proper distance of the farthest galaxy we can see now is

    $$D = cR(t_o) \int_{t_{min}}^{t_o} \frac{dt}{R(t)},$$

    where t_{min} is the time of origin of the universe, t_o is the present time and $R(t)$ is the cosmic scale factor.

28. The distance of an object of known dimensions is calculated from a measurement of the solid angle it subtends, using Euclidean geometry. The result d is obtained. Assuming a uniform model universe, derive a formula for the present proper distance D of the object in terms of d.

29. Prove the result (6.3.7).

30. In Newtonian mechanics, the cosmical constant Λ can be incorporated by adding to gravity an outward radial force on a body of mass m, distant r from the origin, of $F = +m\Lambda r/3$. Assuming that $\Lambda = -10^{-20}$ years^{-2}, estimate the maximum speed a body will attain if its orbit is comparable in size with the solar system ($\sim \frac{1}{2}$ a light-day) (assume F is the only force acting).

31. Calculate the present proper distance D for (a) the object horizon, (b) the event horizon, on the steady state model.

32. Show that the steady state model does not predict that the night sky is ablaze with light.

33. If G changes slowly with time, show that the radial distance from the Sun of a planet in circular orbit varies as G^{-1}. (Hint: what is conserved as G varies?)

34. How does the matter temperature T_m vary with the scale factor $R(t)$ in the matter-dominated era of a uniform model universe?

Solutions to odd-numbered problems

1. Let the masses of the Earth, Mars, and the Sun be $m_\oplus$, m_δ and $m_\odot$, and let the distance of Mars from the Sun be r_δ. Then

$$r_\delta = r_\oplus + R.$$

For circular orbits, Newton's second law gives, for the Earth,

Force $= Gm_\odot\, m_\oplus/r_\oplus^2 = m_\oplus \times$ inward acceleration

$$= m_\oplus \times (\text{velocity})^2/r_\oplus \equiv (m_\oplus/r_\oplus)\,(2\pi r_\oplus/T_\oplus)^2,$$

i.e.

$$Gm_\odot/r_\oplus^{\,3} = 4\pi^2/T_\oplus^2.$$

For Mars,

$$Gm_\odot/r_\delta^3 = 4\pi^2/T_\delta^2.$$

Therefore $r_\delta = r_\oplus(T_\delta/T_\oplus)^{\frac{2}{3}}$

▶ and $r_\oplus = R/[(T_\delta/T_\oplus)^{\frac{2}{3}} - 1]$.

3. M is the *apparent* magnitude the star would have if it were 10 pc away. The apparent luminosity l it would have at this distance is given by

$$l/l_{ref} = 100^{(m_{ref}-M)/5}, \qquad\qquad \text{(equation (2.2.6))}$$

where l_{ref} is the apparent luminosity corresponding to a reference magnitude m_{ref}. Any m_{ref} will do, and we take $m_{ref} = 0$; then we know that l_{ref} is defined as

$$l_{ref} = 2.52 \times 10^{-8}\ \text{W m}^{-2}. \qquad\qquad \text{(equation (2.2.7))}$$

Therefore $l = 2.52 \times 10^{-8} \times 100^{-M/5}\ \text{W m}^{-2}$.

To find L we use the inverse-square law (2.2.5); this gives

$$L = l \times 4\pi \times (10\ \text{pc in metres})^2$$
$$= 2.52 \times 10^{-8} \times 100^{-M/5} \times 4\pi \times (10 \times 3.26 \times 9.46 \times 10^{15})^2\ \text{W}.$$

▶ Therefore $L = 3.06 \times 10^{28} \times 100^{-M/5}\ \text{W}$.

5. *Outside* the central core, stars orbit under the influence of the whole galactic mass M_g, thus Newton's second law gives

▶ $v^2/r = GM/r^2$, therefore $v \propto r^{-\frac{1}{2}}$.

Inside, stars only feel the gravitational attraction of mass closer to the centre than they are. This mass is $M \propto r^3$, so

▶ $v^2/r \propto Gr^3/r^2$, therefore $v \propto r$.

7. Consider the expansion of a cube of side L. In time Δt, each side expands by $LH\Delta t$; this follows from Hubble's law, and H is Hubble's constant. Thus the new volume is

$$(L+LH\Delta t)^3 \approx L^3 + 3L^3 H\Delta t.$$

The rate of increase of any volume V is therefore

$$\frac{dV}{dt} = 3HV.$$

If no matter were created, the rate of decrease of density ρ would be

$$-\frac{d\rho}{dt} = 3H\rho, \quad \text{since } \rho \propto 1/V.$$

To keep ρ constant, matter must be created at the rate $3H\rho$ kg m^{-3} s^{-1}. If M_H is the mass of a hydrogen atom, this is equivalent to

$(3H\rho/M_H)$ atoms m^{-3} s^{-1} $\approx 5 \times 10^{-19}$ atoms m^{-3} s^{-1}.

Thus in every cubic metre, one atom should be created every

▶ $\frac{1}{5} \times 10^{19} = 2 \times 10^{18}$ s $= 10^{11}$ years.

9. From Newtonian mechanics, the period T is

$$T = 2\pi\sqrt{(l/g)},$$

where g is the acceleration of the mass, due to gravity at the place where the pendulum is situated. From the second law, this is given by

$$m_I g = Gm_g M_g/r^2.$$

Thus

▶ $T = 2\pi r \sqrt{(lm_I/m_g M_g G)}.$

11. By assumption, $F \propto ma$. However, the acting mass M must be involved as well, and presumably r and the natural constants G and c. The simplest formula involves these quantities as simple powers, and we write

$$F = Kma M^h r^i G^j c^k,$$

where K is a dimensionless constant. If μ, L, t denote dimensions of mass, length and time, the dimensions of the equation read

$$\mu L t^{-2} = \mu L t^{-2} \mu^h L^i (L^3 t^{-2} \mu^{-1})^j (Lt^{-1})^k$$

i.e. $h-j = 0$, $i+3j+k = 0$, $-2j-k = 0$.

These three equations cannot determine the four unknowns h, i, j, k but we notice that m and M must occur symmetrically in F in order to satisfy the third law. Thus $h = 1$. Then the equations give

$$j = 1, \quad k = -2, \quad i = -1.$$

The simplest choice for K is $+1$, so the final 'simplest possible' formula is

▶ $F = GMma/c^2 r.$

13. As the number might have suggested, this is a trick question. Actually the light is *farther* from the Earth after travelling 1 km. Assume first

that the light travels in a straight line for a distance *l*. By simple geometry, the height *h* through which it is raised is

$$h \approx l^2/2R_\oplus \approx l^2/(2 \times 6000) \text{ km} = \tfrac{1}{12} \text{ m.}$$

Now we must estimate the modification of this value due to the fact that actually the path curves inward under gravity. Assuming Newtonian mechanics, the amount of this fall is

$$\tfrac{1}{2}g(\text{time})^2 = \tfrac{1}{2}gl^2/c^2 \approx 5 \times 10^{-11} \text{ m!}$$

This is completely negligible compared with $\tfrac{1}{12}$ m, as is the corresponding value from general relativity.

15. Since the events are 'neighbouring', we can consider them in a single local inertial frame, and use special relativity. Since the separation is spacelike, it is always possible to choose a frame in which the events are simultaneous. Thus we can write the coordinates of the two events, which we call *O* and *P*, as

$$O: x_O^i = (0, 0, 0, 0)$$
$$P: x_P^i = (0, D, 0, 0).$$

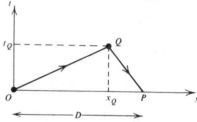

Along the straight world line, the proper distance is

$$D_{OP} = \sqrt{(-c^2\Delta\tau^2)} = \sqrt{\{-c^2[\Delta t^2 - (\Delta x^2 + \Delta y^2 + \Delta z^2)/c^2]\}}$$
$$= \sqrt{[-c^2(0 - D^2/c^2)]} = D,$$

which is obvious.

Now consider the world line *OQP*, where, for *Q*,

$$x_Q^i = (t_Q, x_Q, 0, 0);$$

we have

$$D_{OQP} = D_{OQ} + D_{QP} = \sqrt{[-c^2(t_Q^2 - x_Q^2/c^2)]} + \sqrt{\{-c^2[t_Q^2 - (D - x_Q)^2/c^2]\}}$$
$$= \sqrt{(x_Q^2 - c^2 t_Q^2)} + \sqrt{[(D - x_Q)^2 - c^2 t_Q^2]}.$$

But

$$D_{OP} = D = x_Q + D - x_Q > D_{OQP}.$$

Now any path from *O* to *P* can be made up of segments like *OQ* and *QP*, so *OP* has the *longest* proper distance.   Q.E.D.

17. For any triangle on a sphere with angles α, β, γ there is always another triangle with angles

$$\alpha' = 360° - \alpha, \quad \beta' = 360° - \beta, \quad \gamma' = 360° - \gamma,$$

whose area consists of that part of the sphere not occupied by the first triangle. Thus the angle sum of both triangles is

$$\alpha' + \alpha + \beta' + \beta + \gamma' + \gamma = 3 \times 360° = 1080°.$$

Because the curvature is positive, the angle sum of a single triangle must exceed 180° (the value is the limit for infinitesimally small triangles).

Therefore $\quad \alpha + \beta + \gamma > 180°$

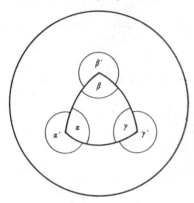

and

$$\alpha' + \beta' + \gamma' < 1080° - 180° = 900°.$$

Therefore, for any triangle on a sphere,

▶ $\quad 180° < \alpha + \beta + \gamma < 900°.$

(Actually, this result holds on any positively-curved surface.)

19. The formula is

$$K = -\frac{1}{2g_{11}g_{22}}\left(\frac{\partial^2 g_{11}}{(\partial x^2)^2} + \frac{\partial^2 g_{22}}{(dx^1)^2}\right) + \frac{1}{4g_{11}^2 g_{22}}\left(\frac{\partial g_{11}}{\partial x^1}\frac{\partial g_{22}}{\partial x^1} + \left(\frac{\partial g_{11}}{\partial x^2}\right)^2\right)$$

[Term no] $\qquad\qquad\qquad 1 \qquad\qquad 2 \qquad\qquad\qquad\qquad 3 \qquad\qquad\qquad 4$

$$+ \frac{1}{4g_{11}g_{22}^2}\left(\frac{\partial g_{22}}{\partial x^2}\frac{\partial g_{11}}{\partial x^2} + \left(\frac{\partial g_{22}}{\partial x^1}\right)^2\right).$$

$$\qquad\qquad\qquad\qquad 5 \qquad\qquad\qquad 6$$

For a plane, the distance formula in polar coordinates is

$\Delta s^2 = \Delta r^2 + r^2 \Delta\phi^2$, i.e. $g_{11} = 1, g_{22} = r^2 = (x^1)^2$

(we use coordinates $x^1 = r$, $x^2 = \phi$). Then all terms in K are zero except:

term 2: $\quad -\dfrac{1}{2g_{11}g_{22}}\dfrac{\partial^2 g_{22}}{\partial(x^1)^2} = -\dfrac{1}{2(x^1)^2}2 = -\dfrac{1}{(x^1)^2}$

term 6: $\dfrac{1}{4g_{11}g_{22}^2}\left(\dfrac{\partial g_{22}}{\partial x^1}\right)^2 = \dfrac{1}{4(x^1)^4}(2x^1)^2 = \dfrac{1}{(x^1)^2}.$

▶ Therefore $K = 0$ Q.E.D.

For a spherical surface, the distance formula is

$\Delta s^2 = R^2\Delta\theta^2 + R^2\sin^2\theta\,\Delta\phi^2$, i.e. $g_{11} = R^2$, $g_{22} = R^2\sin^2\theta \equiv R^2\sin^2 x^1$

(we use coordinates $x^1 = \theta$, $x^2 = \phi$). Then all terms in K are zero except:

Term 2: $-\dfrac{1}{2g_{11}g_{22}}\dfrac{\partial^2 g_{22}}{(\partial x^1)^2} = -\dfrac{R^2 2\cos 2x^1}{2R^4\sin^2 x^1} = -\dfrac{\cot^2 x^1}{R^2} + \dfrac{1}{R^2}$

Term 6: $\dfrac{1}{4g_{11}g_{22}^2}\left(\dfrac{\partial g_{22}}{\partial x^1}\right)^2 = \dfrac{R^4 4\sin^2 x^1\cos^2 x^1}{4R^6\sin^4 x^1} = \dfrac{\cot^2 x^1}{R^2}.$

▶   Therefore $K = 1/R^2$. Q.E.D.

21. The metric distance formula for a space of constant curvature K in polar coordinates r, θ, ϕ is

$\Delta s^2 = \Delta r^2/(1 - Kr^2) + r^2\Delta\theta^2 + r^2\sin^2\theta\,\Delta\phi^2$.

For the r-sphere, the circumference around the 'equatorial circle' $\theta = \pi/2$ is

$$C = \int_{\substack{\Delta r=\Delta\theta=0\\ \theta=\pi/2}} \mathrm{d}s = \int_0^{2\pi} r\,\mathrm{d}\phi = 2\pi r.$$

The proper radius is

$$a = \int_{\Delta\theta=\Delta\phi=0} \mathrm{d}s = \int_0^r \dfrac{\mathrm{d}r'}{\sqrt{(1 - K(r')^2)}} = \dfrac{1}{\sqrt{K}}\arcsin r\sqrt{K},$$

i.e. $r = (1/\sqrt{K})\sin a\sqrt{K}$.

▶   Therefore $C = (2\pi/\sqrt{K})\sin a\sqrt{K}$.

(If K or a tends to zero $C \to 2\pi a$ as expected.) The area of the r-sphere is $A = 4\pi r^2$ by definition, i.e.

$A = 4\pi\sin^2 a\sqrt{K}/K$.

The volume V is the sum of volumes of successive shells,

i.e. $V = \displaystyle\int_0^a A(a')\,\mathrm{d}a' = \dfrac{4\pi}{K}\int_0^a \mathrm{d}a'\sin^2 a'\sqrt{K}$

$$= \dfrac{2\pi}{K}\int_0^a \mathrm{d}a'\,(1 - \cos 2a'\sqrt{K})$$

▶   i.e. $V = \dfrac{2\pi a}{K}\left(1 - \dfrac{\sin 2a\sqrt{K}}{2a\sqrt{K}}\right)$.

(If K or a tends to zero, $V \to \frac{4}{3}\pi a^3$ as expected.)

23. The Schwarzschild radius r_s is defined as

$r_s = 2GM/c^2$.

For a galaxy, $M \sim 10^{11}M_\odot$, so

▶ $r_s \sim 10^{11}r_{s\odot} \sim 3\times 10^{14}$ m $\sim 10^{-2}$ pc.

For a proton, $M = 1.7\times 10^{-27}$ kg, so

▶ $r_s = 2.6\times 10^{-54}$ m.

25. *Case (a)*. In Newtonian mechanics, for a body with speed c (light) to move in a circular orbit about a mass M, the radius r must satisfy
Inward acceleration $=$ mass $\times c^2/r =$ force $=$ mass $\times MG/r^2$,

▶ i.e. $r = GM/c^2 = r_s/2$.

Case (b). In general relativity, we use the orbit equation (5.4.2.) for a null geodesic; this is

$$D = \frac{(dr/d\phi)^2}{r^4(\phi)} + \frac{1}{r^2(\phi)} - \frac{2GM}{c^2 r^3(\phi)}.$$

Differentiating with respect to ϕ gives, after cancelling a common factor $dr/d\phi$,

$$0 = \frac{2d^2r/d\phi^2}{r^4} - \frac{4(dr/d\phi)^2}{r^5} - \frac{2}{r^3} + \frac{6GM}{c^2 r^4}.$$

For a circular orbit, r is constant, so
$2/r^3 = 6GM/c^2 r^4$,

▶ i.e. $r = 3GM/c^2 = 3r_s/2$.

27. The farthest galaxy we can see now has comoving radial coordinate σ_{oh} lying at the *object horizon*. From (6.2.6) this is given by

$$\int_{t_{min}}^{t_o} \frac{dt}{R(t)} = \frac{1}{c} \int_0^{\sigma_{oh}} \frac{d\sigma}{\sqrt{(1-k\sigma^2)}}.$$

But from (6.1.4) the present proper distance to σ_{oh} is

$$D = R(t_o) \int_0^{\sigma_{oh}} \frac{d\sigma}{\sqrt{(1-k\sigma^2)}}.$$

▶ Therefore $\quad D = cR(t_o) \int_{t_{min}}^{t_o} \frac{dt}{R(t)}.$ Q.E.D.

29. We start from (6.3.6), namely

$$\frac{l}{L} = \frac{H_o^2}{4\pi c^2 z^2} [1 + (q_o^2 - 1) z + \dots].$$

L is the absolute luminosity; the apparent luminosity the galaxy would have at 10 pc is l_{10}, given by

$l_{10} = L/4\pi(10 \text{ pc})^2.$

Therefore $\quad \dfrac{l}{l_{10}} = \dfrac{H_o^2 (10 \text{ pc})^2}{c^2 z^2} [1 + (q_o - 1) z + \dots] = 100^{(M-m)/5},$

where the last equality follows from (2.2.6). Taking logs, we get

$$M - m = \tfrac{5}{2} \log_{10} \left\{ \frac{H_o^2 (10 \text{ pc})^2}{c^2 z^2} [1 + (q_o - 1) z + \dots] \right\}$$

$$= 5 \log_{10} H_o - 5 \log_{10} cz + 5 \log_{10} (10 \text{ pc})$$

$$+ \frac{5 \log_e [1 + (q_o - 1) z]}{2 \log_e 10} \dots.$$

Now

$$\log_{10} (10 \text{ pc} \times H_o) = \log_{10} (H_o(\text{km s}^{-1} \text{ Mpc}^{-1}) \, 10^{-6} \times 10 \text{ pc})$$
$$= -5 + \log_{10} (H_o(\text{km s}^{-1} \text{ Mpc}^{-1})).$$

Therefore

▶ $$M - m = 5 \log_{10} (H_o \, (\text{km s}^{-1} \text{ Mpc}^{-1})) - 25 - 5 \log_{10} cz + 1.086(q_o - 1) \, z.$$
Q.E.D.

31. For the steady state model, the cosmic scale factor is

$$R(t) = A \exp (H_o t). \tag{equation (7.3.1)}$$

The space is flat, so that the proper distance D corresponding to a co-moving coordinate σ is, at the present time t_o:

$$D = R(t_o) \, \sigma.$$

This universe does not explode or implode, so it has an infinite past and future. From (6.2.6) the coordinate σ_{oh} of the *object horizon* is

$$\sigma_{oh} = c \int_{-\infty}^{t_o} \frac{dt}{R(t)} = \frac{c}{A} \int_{-\infty}^{t_o} dt \exp (-H_o t) = \infty.$$

▶ Therefore $D = \infty$.

There is no object horizon: all objects can in principle be seen by us. From (6.2.7), the coordinate σ_{eh} of the *event horizon* is

$$\sigma_{eh} = c \int_{t_o}^{\infty} \frac{dt}{R(t)} = \frac{c}{A} \int_{t_o}^{\infty} dt \exp (-H_o t) = \frac{c \exp (-H_o t_o)}{H_o A} = \frac{c}{H_o R(t_o)}.$$

▶ Therefore $D = c/H_o$.

Events occurring beyond this proper distance at the present cosmic time will never be seen by us.

33. Let the planet move in a circular orbit of radius $r(t)$ with speed $v(t)$, at time t, when the gravitational 'constant' is $G(t)$. Then Newton's second law gives

$$v^2(t)/r(t) = G(t) \, M_\odot/r^2(t).$$

Throughout the motion, angular momentum (L) is conserved, since no torque acts on the planet. Therefore

$$L = v(t) r(t) = \text{const.},$$

and $$L^2/r^3(t) = G(t) \, M_\odot/r^2(t),$$

or

▶ $$r(t) = L^2/M_\odot \, G(t) \propto G^{-1}. \quad \text{Q.E.D.}$$

Useful numbers

Speed of light c: 2.998×10^8 m s^{-1}

Gravitational constant G: 6.67×10^{-11} m^3 kg^{-1} s^{-2}

One light-year: 9.46×10^{15} m $= 0.307$ pc

One parsec: 3.09×10^{16} m $= 3.26$ light-years

One year: 3.156×10^7 s

Mass of Sun $M\odot$ 1.99×10^{30} kg

Radius of Sun $R\odot$ 6.96×10^8 m

Mass of Earth $M_\oplus$ 5.98×10^{24} kg

Radius of Earth $R_\oplus$ (mean): 6.37×10^6 m

Hubble's constant H_0: (55 ± 7) km s^{-1} Mpc^{-1}

$$= (1.8 \times 10^{10} \text{ years})^{-1}$$

Deceleration parameter (estimated) q_0: 1 ± 1

Present mass density of universe (estimated) ρ_0: 3×10^{-28} kg m^{-3}

Bibliography

French, A. P. *Newtonian mechanics* (Norton, New York 1971). Extensive treatment at undergraduate level, clearly presented and full of historical insights.

Sandage, A. (ed.). *The hubble atlas of galaxies* (Carnegie Institution, Washington 1961). A beautiful collection of photographs showing the variety of forms taken by galaxies.

Misner, C. W., Thorne, K. S. and Wheeler, J. A. *Gravitation* (Freeman, San Francisco 1974). Massive graduate-level text on general relativity, that will undoubtedly become a classic. Fascinating to dip into for historical and philosophical perspectives, and speculations about the deepest unsolved problems in physics.

Rindler, W. *Essential relativity* (Van Nostrand Reinhold, New York 1969). Very clear discussion of special relativity, and the simplest treatment of general relativity employing tensors.

Sciama, D. W. *Modern cosmology* (Cambridge University Press, Cambridge 1971). A largely descriptive account at the undergraduate level, by a well-known cosmologist. Extensive discussions of the astrophysical and radio-astronomical aspects of the subject.

Synge, J. L. and Schild, A. *Tensor calculus* (University of Toronto Press 1959). The best treatment of this difficult branch of mathematics.

Weinberg, S. *Gravitation and cosmology: principles and applications of the general theory of relativity* (Wiley, New York 1972). Advanced text by a theoretical physicist famous for his work on elementary particles. An individualistic treatment full of interesting insights.

Index